Design and Build Contract Practice

Second Edition

Dennis F. Turner

BA(Hons), FRICS, MCIOB

Longman Scientific & Technical
Longman Group Limited
Longman House, Burnt Mill, Harlow
Essex CM20 2JE, England
and Associated Companies throughout the world

First published 1986
Second edition 1995

British Library Cataloguing in Publication Data
A catalogue entry for this title is available from the British Library

ISBN 0-582-08968-9

Set by 3 in 10/12 Plantin
Printed and bound in Great Britain
at Bookcraft (Bath) Ltd.

Contents

Part 2 Practice and the JCT contract

Part 3 Alternative approaches

Introduction

The advent of a second edition always causes mixed reactions, at least in this author. It means that the concept of the book is under review and that the resulting revision is going to disturb what was intended to be a reasonably logical pattern of ideas – did I get it right, why pull it about, and am I just getting set in my ways? The countervailing reaction must be some gratification, mixed with relief, that the first edition must have been useful somewhere, to someone, somehow. One may be left wondering what went right and what wrong. In this instance in the course of consultancy and lecturing, I have had enough comments of various hues to feel encouraged about the direction of revisions. Indeed, the subject of design and build appears to have been one of steadily growing interest over the time since the first edition was published.

In the original introduction I noted that the topic of design and build had not received a great measure of overall treatment, despite the existence of its own contract documents, published articles, lectures and other available treatment. Each of these sectors has expanded considerably to give a more rounded view of the whole process, and numbers of us have made contributions accordingly. It still looks as though the present book has a special part to play by giving that overall treatment which should enable newcomers, whether students or practitioners from related areas, to secure a frame of reference, and afford those of varying degrees of experience in the field the chance to check out their methods and, hopefully, gain additional insights on the way. These should include the obvious participants, clients, contractors and design and cost consultants, but also specialists in the development, legal and financial professions who are increasingly involved in projects today. Students are especially likely to be those engaged on courses which encourage enquiry and a comparative understanding of aspects which cross traditional subject boundaries.

Since the previous edition, quite a number of developments have occurred or become more prominent in the wider context for design and build and inside its methodology and these have led to expanded or additional treatment of topics.

There are more sophisticated clients who have insisted on procedural and contractual innovations in their favour, with varying success in the outcomes, including novation of consultants and collateral warranties afforded by just about everyone to everyone else, but especially to clients and later owners. Numbers of contractors have become more established in the market, which itself has grown so that some estimates (always hard to form) suggest that in the industrial and commercial sectors design and build is approaching half the market share in value as a procurement method. Some at least have produced work of a standard of design which compares well with that produced by other means. Consultants are entering the field for at least two reasons: on the one hand, the value of the system is being appreciated in suitable cases. On the other work has become scarcer in total and this has been accompanied by the proportionate increase in design and build, so that it is a case of adapting to a changing market.

In methodology, it is now necessary to deal with the particular emphasis of professional indemnity insurance in the field and to take note of the more intricate liabilities of consultants as their involvement can span different stages of projects and come from different directions. The consideration of tendering has been expanded by treatment of how the tendering activities of contractors are affected when design is included. Numbers of the practices introduced by the more independently minded employers, and points for their caution, are taken in passing – while novation and collateral warranties have separate treatment.

As ever, the contractual mechanisms have not remained static. The JCT contract has seen substantial amendment in the areas of insurance and determination, with numerous smaller changes (such as over quality and payment) which in total are significant. Over the horizon is the promise, subject to indefinite extension of time, of 'section headed format' which may drastically shuffle not the meaning, but the standard arrangement of cards (otherwise clauses) in the JCT pack one day. This has justified some preliminary treatment in an appendix to ease the transition.

Importantly, there have been additions to the contractual modes of dealing with design and build, or its equivalents, now available. Within the JCT range, there is the addition of a supplement to the with-contractor's design form to relate to the BPF system and of performance specified work clauses in the JCT standard form. Elsewhere, a design and build version of the latest GC/Works/1 contracts and a design and construct version of the latest ICE contract have appeared. All of these have merited a chapter apiece, even though one does not relate to building work. This has expanded the comparative treatment approach of Part 3 sufficiently to lead to an introductory chapter to link the quite extensive range of contracts etc together properly. Almost needless to record, legal decisions have not remained unsupplemented in our litigious age.

In the light of all this, I have held to the original structure of the book as serving its purpose, but amplified it where necessary. I therefore essentially repeat here what I wrote before. The resulting structure of this book is based on practice, in that it takes underlying aspects and then runs through the sequence of events that a project encounters. At each stage of the sequence, the related elements of the JCT with contractor's design contract are discussed, as practice cannot shake itself free from its contractual basis at many points. This has been seen as a necessary structure to bring the wholeness of the subject into focus, but it has caused quite a bit of arrangement and rearrangement of even the most straightforward parts to secure the right flow of ideas. There is some repetition of ideas and reference to another side of the discussion somewhere else. This is deliberate, in that certain parts of the argument warrant repetition. There is an old homiletic dictum which goes 'tell them what you are going to say, then tell them it, and then tell them what you have told them'. Perhaps in these more mundane fields it counts as well.

The practice content tends to be by way of discussion, rather than to follow the line of a code of practice. This is because affairs are in a fairly pragmatic state of development, perhaps meaning that behaviour has not yet fossilised. A code of practice has been issued and is welcome as emphasising the essentials of good behaviour. By contrast, the contractual content is more constrained, because it is discussing fossils, as clauses are by nature.

Contractual matters are taken in a self-contained way, so that all areas are treated in the case of the JCT contract, although with the greater emphasis upon the clauses relating more to design and build as such. Other contracts and elements of contracts are taken more selectively to bring out their design and build features, to show how these compare with the JCT approach and to make the survey rounded.

It has been kept in mind that among the intended readers may be those who are 'lay' so far as the peculiarities of building work are concerned, so that what may be obvious to the initiated is laboured in places. The differences in treatment have not been flagged up by special typeface or any such device! They should be fairly clear without breaking up the discussion. As always, the caveat applies that expert advice should be sought in practice. This is true over technical and legal matters alike. To this end, legal cases in the text have been restricted to those necessary to provide some background to the peculiar aspects of design and build, and treatment is fairly summary. The table of cases includes other cases bearing on wider aspects. Quotations from the contracts are highly selective and, as always, the maxim is 'read the original and not just what is said about it'.

As ever, I must acknowledge my debt to the many from whom one acquires that invaluable ingredient of experience which comes from constructive interchange. Over recent years these have included numerous individuals and organisations with whom I have had dealings through consultant and lecturing situations. I would single out my brother, Alan E. Turner, with whom much fat has been chewed on many issues. As I have recorded in the introduction to the revised edition of another book, I greatly appreciate the help and support afforded me by that paragon of patience, my new wife Carol, who has assumed the mantle of my dear late Betty with fortitude, in this respect and probably in many others. Life is never a lone endeavour.

April 1994 Dennis F. Turner

Part 1

Basic aspects

Chapter 1

Client needs and contract solutions

The employer's aims and needs in building

Some contractual alternatives

The design and build alternative

There is a wide range of solutions to the problem of how to arrange contractually for building work to be carried out. The persons most directly affected are the building client (referred to in most chapters in this book as 'the employer') and the contractor. They, after all, respectively either pay for the work or perform it; and their concerns are widely discussed in this volume.

The reasons for so many contractual solutions, lie either in economic trends, in passing or recurring fashions, but also in the nature of building. Leaving aside the speculative housing sector, involving 'contracts of sale', the 'contracts to build' sector embraces new work and alterations on all scales, related to many types of construction. In addition, the employer may be an individual, a partnership or one of a variety of public or private corporate bodies. For some, commissioning building work is a once-in-life-time experience; for others, it is part of or even the whole of their activity,. Thus contractors have developed in various ways to meet the needs of different employers.

A number of contractual approaches have developed in this environment and become 'traditional' over the last century or so; however, in recent years several alternatives have come forward. The evolution of these traditional approaches is now a matter of history; the fact is that they all incorporate the use of professional consultants in a similar way. Architects, engineers, other designers and quantity surveyors have separate contracts with the employer and work in the communications gap between him and the contractor, who has a contract solely with the employer. There is thus a separation of design and its cost control from construction, not only because there are distinct specialists in each, as is almost inevitable, but because the functions are performed by different organisations which exercise limited control over one another. The strengths and weaknesses of this structure are mentioned in Chapter 2.

Among other things, these newer alternatives usually seek to amend the communications structure; thus project management either introduces a further consultant or elevates an existing one to co-ordinate the activities of everyone else.

Some would say that the architect has traditionally fulfilled this task and is well placed to continue doing so. Management contracting puts the contractor into the position of project management, while retaining the individual consultants. However, neither of these alternatives is taken further in this book, which is about the more radical alternative of 'design and build', as used for the whole or part of a project.

Design and build is the umbrella term also covering package contracting, the all-in service, develop and construct, and turnkey contracting. Package is a shorthand for design and build, meaning the performance of both functions in one 'package', that is by the contractor as a single contractual person; however, he may parcel out the work to his own consultants. It usually carries the connotation of system or industrialised building, such that the project can largely be selected from a catalogue or by viewing an already existing example. All-in service is a direct, if vague, equivalent of design and build. Develop and construct requires some design work by the employer or his consultants, which the contractor takes over and 'develops'. Any scheme, of course, requires some briefing from the employer, even if it is only the provision of a schedule of accommodation resulting from a minimal amount of briefing. It is thus at one end of the spectrum of design and build. The idea of a turnkey contract at its most embracing is that the contractor acquires the site and secures all approvals for the client in addition

Table 1.1 Building contract relationships

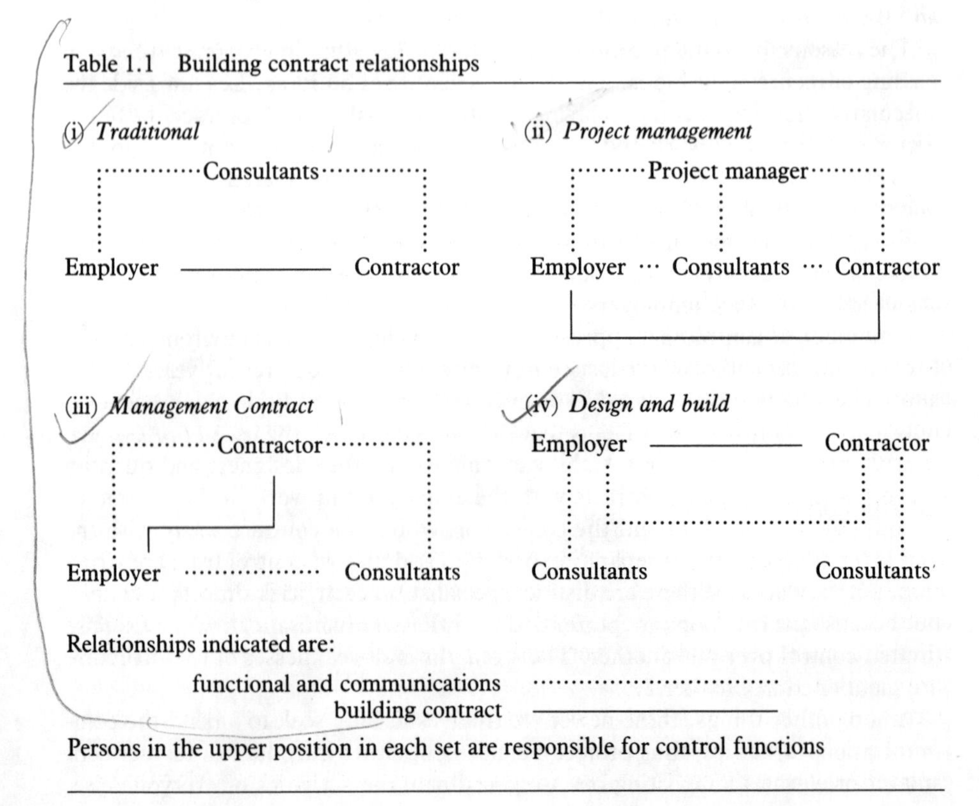

to designing and building and installing all plant, equipment and even furnishings. The employer simply has to 'turn the key', walk in and start using the place. These latter two vary the design and build theme respectively at the beginning and the end, without destroying it. While they are not referred to explicitly, they are accommodated within design and build, which is widely used hereafter.

The effect of design and build is to move the consultants from the gap position between employer and contractor and place them under the sole authority, and possibly within the permanent organisation, of the contractor, whom they alone advise. As a counteraction, the employer may then engage further consultants to give him independent advice on some issues. This is similar to the pattern of many other industries, which do not separate design and production in the way that construction does. It has its strengths and weaknesses, which are mentioned later in this chapter, and will become apparent in further discussion.

The essence of the distinctions drawn so far is presented diagrammatically in Table 1.1. While commercial forces do in time lead to changes and while fashions rise, fall and revive, it is not unreasonable to suggest that there is no optimum procurement pattern at any one time. Within this diversity, some (by no means all) employers are looking for simplicity and economy, while, to satisfy them, some contractors are promoting design and build.

In the main, this book refers to employers and contractors, who are taken to include their various consultants to avoid constant repetition. The special positions of consultants do raise some distinct issues, which are treated in Chapter 3.

The employer's aims and needs in building

Different contractual solutions have different effects in furthering the employer's aims to meet his needs; they regulate various aspects of the bargain with differing precision. Since not all employers' aims are the same they are best served by these various solutions. In general the employer states what he wants, where and when, together with any budget limits; the precise detail is usually worked out by others. Aims like those set out below tend to influence where the boundaries are drawn. This section outlines the more prevalent aims which may be recalled in later discussions, where some but not all are referred to specifically.

Satisfaction of function

This is the primary aim for any employer, in the sense that unless he has certain needs to be met, and unless the proposed building meets them, he will not commission it at all. This is true whether he wishes to use or sell the product and whether, say, it is intended for personal enjoyment, commerce or industry. The simplicity or complexity of function may vary widely, whether the building's purpose is intended to be static or change during its life: in all cases the employer's needs must be met. This also is a truism, but worth this brief emphasis especially concerning concepts like design, performance and fitness for purpose.

Contracts and the law do not always meet 'common sense' expectations fully on these points.

Function, taken alone, is concerned at the macro-level with design which provides facilities which work well, assuming adequate materials and workmanship are always provided. This therefore needs specialised design, ahead of construction. Among those who supply design are, for example, a generalised specialist in design, such as an architect or structural engineer who may have wide competence and insight; or a specialist in a particularly narrow area of design, such as a refrigeration engineer, with his depth of knowledge and, possibly, construction experience. There are ready arguments for and against each of these performing particular design work, although there appears little in favour of the specialist in construction work who is not also a regular designer, especially when providing the broad scheme design.

Achievement of economy

This means gaining the best result, which may not necessarily be the cheapest, for what the employer is prepared to spend. 'Best result' begs a lot of questions, that must be answered by the other aims. At the lowest level, if the employer cannot afford it at all he cannot have it, whatever the function, aesthetic or economic return. Three sequential areas of economy may be instanced which overlap each other.

The first is strategic and less a matter of building design: where to locate the building geographically in relation to site characteristics. This may be fixed by prior ownership of a site, or there may be a range of sites available with a programme of building works phased over years. The advice that the employer needs may come from within his own organisation or need special consultancy. This precedes and is distinct from design, whoever performs it.

The second area is design itself where, within the context of 'value for money', there is need to control the initial costs. These are the amounts paid for the design and the construction, and the costs of financing them, including the loss of use of money owned or borrowed which is paid out on account and on which there is no return until the building is in use. The concerns of employer, designer and constructor must all be met, for it is the employer who actually pays out; the designer has his design liability to protect; and the constructor has a profit to make. Separate cost advice on the design and control of what is paid is a way of limiting the problems.

The third area of economy is still design, but in relation to controlling the continuing costs after construction, and so the total or life cycle costs. These include operating the building (particularly staffing and energy), cleaning, maintenance and repair, alteration and adaptation, and depreciation to the point of sale or demolition, with disposal costs. Again, the concerns of the various people involved must be given consideration, and this may be complicated by any tension between initial and continuing costs. These should be resolved by a properly discounted calculation of total costs. Unless the employer is given the full picture for the life of the building, he may be misled as to the efficiency of the design or designs which are offered in a set of tender figures. This is especially the case when contractors are in competition over design as well as construction.

Within all these factors the level of the tender and the final settlement must be considered. This is influenced by the economic climate, the degree of competition and the contract arrangement used. Some of these control expenditure more tightly or at least more predictably than others. Design and build has the advantage that both its functions are within or controlled by one organisation, which may be a potential factor for economy, although an incentive to economise is needed.

Control of programme time

For some employers, this may be the tail that wags the dog strongly, or even irresistibly takes control. Properly considered, it is a sub-set of economy, often viewed in relation to the construction period alone, when the employer is paying out his largest sums of money and the contractor has continuous site overhead costs. These elements have to be balanced against the efficient purchase and use of the other factors of production on and off site, which in turn affect the contractor's costs allowed in his tender. The whole programme should be considered from the first clear articulation of the need for the project to the time at which it might come into use. This should be viewed in relation to the subsidiary optimum time for total production economy. The need to meet a fixed date for public or commercial reasons may be extremely strong, or funds may be available only in a limited flow, so compressing or extending this time.

Following on from the above, there would be an apportionment within the whole programme of the various elements. Whoever does what, there must be a brief supplied as ideas are clarified; development of a design and entering into a contract; and construction and handing-over. These are often regarded as discrete and sequential activities, and certainly there must be clarity as which is to be performed and when, to give a balanced programme. Almost invariably these activities, while distinct in concept, overlap in time if only because a project can be broken into smaller parts proceeding at various speeds. These same activities may be subject to reiteration as design throws up questions. In addition to this built-in, cyclical effect, there may be a conscious overlapping of the activities to achieve compression of the overall programme. This may work well, so long as reiteration does not then undermine the work of later stages with too many changes.

All contract arrangements, by definition, give an absolute demarcation at the stage of entering into the contract: some require complete design before tendering, some allow design to continue during construction, although ahead of the section concerned! Most lump sum contracts require the separation of the two, contracts based on working out the price payable as work proceeds allow an overlap of design and construction. Design and build contracts are lump sum contracts, but allow design to proceed during construction usually without adjustment of the price. Quite commonly, established contractors in the field will quote a guaranteed maximum price, with any saving shared with the employer. These cases depend on reimbursement methods which are discussed briefly later in this chapter. Design and build contracts as a class are not so flexible as some arrangements as the contractor must be reasonably sure of his design before committing himself irrevocably. However, the contractor does have control of both design and construction and so should be able to integrate the work closely.

Creation of aesthetic conditions

This title is no doubt an exercise in understatement and relates to the visual and environmental qualities of a building which defy objective analysis when present; though when they are absent it is only too obvious. They are also not so directly susceptible to monetary evaluation as the preceding aims although, if present, are worth something when selling. They add to the occupants' and to society's well-being; after all, more people pass by buildings than enter them.

What actually results must be affected to some extent by the function, budget and maybe even programme of the building. But beyond all these is that extra 'something' in the good designer. It is sometimes argued that design and build schemes are not so high in this respect as those where the design function is not under the same commercial umbrella and constraint. No doubt examples for and against can be produced from both areas of design, going right back to the Middle Ages. This point recurs in later chapters.

Some contractual alternatives

Three distinct alternatives are outlined here and are the main types within what has already been called the traditional group. Each of them is based therefore upon the segregation of design and construction into distinct functions, so that under each contract procedure design is primarily the right and duty of someone, a consultant, other than the constructor.

Usually within the first two alternatives only, standard contract conditions (such as those produced by the JCT and mentioned in various places in this book) make limited provision for design of relatively subsidiary parts of the works to be performed through the contract. This may be by the contractor himself, using such arrangements as a designed portion supplement (see Ch. 12) or performance specified work (see Ch. 17), where he may choose to sub-let the design to sub-contractors who are of the regular variety or who are consultants to him. Alternatively, this design may be by a nominated sub-contractor, standing quite distinct from the contractor so far as design is concerned and dealing in this respect directly with the architect, despite being a sub-contractor (see Ch. 18).

Entirety of contract

Before taking the alternatives, a principle may be introduced. A contract is said to be 'entire' when there is a complete obligation laid on each party to perform his side of the bargain. In the case of building, this means:

(a) The employer is bound to pay the whole contract sum.
(b) The contractor is bound to perform all the work to earn payment; he cannot choose to perform only, say, two-thirds and be paid that proportion of the contract sum.

This is an all-or-nothing arrangement. It includes a lump sum, which is usually taken as a sum established when the contract begins. But it can also be based upon an agreement to pay a sum not fixed in advance, but calculated at the end on the basis of a fully defined formula agreed in advance. The alternatives given below are split between these options. Even a contract with an implied term to pay a fair price may be construed as entire. This is a rigid position, unless express terms of the contract give modified entirety. The more important issues involved and modified by most standard contract forms, including those considered in this book, are:

(a) Work and materials reasonably necessary are included, even if not specifically mentioned.
(b) Payment is not due on an interim basis and not due at all if the contractor abandons the works, unless the employer has prevented completion.
(c) Payment is due in full as soon as work is agreed to be complete.
(d) The contractor is liable to complete the works without further charge, even if they are destroyed during progress and he has to rebuild.
(e) Neither party can vary the works being performed, that is introduce any change into the design or specification, although the employer may have something left out without reduction of the contract sum.

While a design and build contract gives a firm commitment, it usually produces a defined extent of flexibility over (b) to (e) in particular.

Lump sum based on drawings and specification

This arrangement offers the most simple financial basis: a single amount as the contract sum, related to a complete set of drawings down to all details and a thorough specification, prepared before tenders are invited. Design and build is a variant of this arrangement, differing basically in the transfer of the design responsibility to the contractor, who may defer part of it to the post-contract period, so overlapping stages.

The traditional system is intended for projects which are fairly small, say up to £100,000 in value, or those which are somewhat larger, but little more than an uncomplicated shed. There is growing disquiet that contractors are more frequently being obliged to contract for projects beyond these bounds on this basis. It depends upon an absolute segregation of design and construction, not only of those performing them, but also of the various stages: design preceding and construction following tendering.

When the contract is formalised both parties, employer and contractor, know their commitments precisely in terms of what is to be constructed and how much is to be paid. This is usually subject to several provisions, of which the most important are those for reimbursing changes in market costs of labour, materials etc (very often not done in design and build) and for making adjustments for variations, usually design changes instructed during construction. To ease calculation of these variation adjustments, it is common to have a schedule of rates embodied in the contract. This is a list of items in the works, expressed in units, such as square metres, each with a rate or price per unit. There is, however, no indication, of the total quantities embodied in the scheme, just an unquantified list. When a variation occurs, the actual amount of it is measured and priced at the unit rate, so that the contract sum is adjusted accordingly. The schedule of rates is a possible device in design and build contracts and is discussed further in Chapter 6.

The effect of this system is to make the contractor responsible, when tendering, for calculating the price *level* at which he will enter the contract, and also for calculating the *amount* of work that he has to perform. He is held to both aspects contractually, and so may gain the contract by keen pricing or by inadequate assessment of the quantities of work. Under the next alternative, he is responsible only for the price level. He thus has more risk here, but only for a relatively small value of work.

Lump sum based on bills of quantities

In legal principle, this arrangement is similar to that just described, except that the content of the lump sum is shown in bills of quantities. These are a detailed analysis of the project, giving a complete list of items contained, each with its quantity set against it. These bills are prepared by the quantity surveyor from drawings etc supplied by the architect or other designers and are supplied to all tenderers, who are to assume their correctness. They insert a unit price against each item; this, together with the quantity given, produces the total for the item. When all items are totalled, with certain other inclusions, the result is the total of the bills and thus the amount of the tender. To prepare the bills, the quantity surveyor needs the same information as is made available to tenderers under the drawings and specification system. However, the tenderers need only a restricted selection from this information, to gauge the *nature* of the works. The *extent* of the works is derived solely from the bills of quantities.

This system, when embodying firm quantities, is intended for schemes larger or more complicated than those covered by the previous system, but which are still designed by someone other than the contractor, before tenders are invited. Again, it is subject to provisions about market cost changes and variations in design etc. In the latter case, much fuller information in the bills of quantities is available as a basis for adjustment. Furthermore, the quantified analysis produced allows interim payments to the contractor during process to be calculated fairly easily, provided it is assumed than an apportionment of the tendering analysis properly reflects the value to the employer of work partly completed.

A variation on the theme is bills of approximate quantities, which are priced in exactly the same way as bills of firm quantities. These allow tenders to be obtained for work which is partly designed at the time, so that the balance of the design is performed later and overlaps construction. As the work is designed and built, it is completely remeasured, taking account of variations and priced at bill prices etc to give a final account. Legally, this is a lump sum ascertained after the event, as the outcome is theoretically predetermined without recourse to considerations of productive efficiency.

With either bills of quantities system in contrast with the drawings and specification approach, the contractor is responsible when tendering for calculating only the price level, as the amount of work is set out for him. Any errors of quantity or description in the bills are corrected and so lead to an adjustment of the sum payable. The contractor, however, is held to his pricing, be it high or low. He does not include in his tender the cost of producing the quantities for the current job and a share of those for which he is unsuccessful. The employer pays for this separately once only, although he also pays for it being done more elaborately, because of the need to present the information in a form clear to those who have not prepared the quantities.

It is generally held that the cost of independent quantities is justified in the larger projects, as the contractor excludes the costs of preparation and also makes no allowance for the risk of his own errors in so doing. Tenders based on approximate quantities tend to be somewhat higher, as there may be some uncertainties in assessing the prices. There is also less certainty for the employer over what he will eventually pay, even though various forecasting techniques are available. They do, however, allow an earlier start on site, although not always earlier completion. Any opportunity for the employer and his consultants to delay design decisions can lead to procrastination which may extend the end-date and increase the final payment!

While quantities in the form and with the contractual status described are not usually suitable even as part of a design and build contract, measured analysis at some point is almost inevitable. The advantages and disadvantages are considered in Chapter 9.

The firm commitment and control in the various approaches so far described suit cases when the extent of the work and the prices can be estimated in advance, so that the contractor is paid for *output* to the employer. This contrasts with the greater uncertainty described next.

Prime cost reimbursement

In cases in which the nature, circumstances or sequence of the work is substantially uncertain, so that the contractor cannot reasonably assess his costs in advance, payment to him for his *input* rather than his *output* may be made by this contract arrangement. It is sometimes known as a cost-plus contract. The essence is for the contractor to be paid for hours of labour and plant and quantities of material as incurred at rates and percentages which allow for their actual or prime cost and certain overheads, while other overheads and profit are paid as a lump sum or on some sliding scale. This removes most of the risk from the contractor, while somewhat limiting his rewards.

Tendering is limited to establishing base rates, percentages and lump sums. What may be more critical than the apparently competitive rates is the contractor's efficiency since, within limits, he is paid for what he expends. So he is often appointed on the basis of a single selection and negotiation of terms, followed by a contract. It is possible when the value of work is fairly clear to build elements into the formula which give the contractor a direct incentive to control costs. A lump sum overheads and profit allowance goes so far, because the larger the prime cost the less in percentage terms does the allowance become. More drastic target versions are based on an agreed estimated cost, with the actual difference shared between the parties or even, in the case of an excess, with it all borne by the contractor. When the anticipated cost is quite uncertain, no such provisions may be possible. There is usually a clause giving the architect some control over whether the contractor is overmanning the site and so forth.

This type of contract is not normally used without particular need, although some employers and contractors enter into it regularly because they have developed a mutual trust over the years. Otherwise it is restricted, as indicated, to cases of uncertainty that make tendering on a lump sum basis impracticable. When just quantity is uncertain, approximate quantities or even a schedule of rates without quantities may be preferable. Prime cost suits cases of intricate alterations, repairs after damage or very urgent work with a confused programme subject to unpredictable constraints or disturbance. There is no inherent limit to the scale of project that may be dealt with in this way.

Such a basis of reimbursement does not go with a plain design and build system (and certainly not one under the JCT form), as it would either amount to a blank cheque or, if strictly controlled, destroy the contractor's initiative and take away his responsibility for his design. It is normal to include a schedule of daywork rates in a design and build contract to be used in valuing self-contained pieces of additional work that cannot be otherwise valued (see Ch. 9). It is fairly common, however, to base a design and build contract on a guaranteed maximum lump sum price, with any reduction of the price calculated by reference to the prime cost and the benefit shared between the parties. This particular approach is not considered in any detail in the present book, although many of the principles given hereafter are as relevant to such a contract, including those most closely related to the design function.

The design and build alternative

This has come into prominence as a drawings and specification contract with designing and specifying as well as construction by the contractor, and with the contract price as a lump sum. It has always been available for works like minor extensions or alterations to houses and maintenance while the specialist subcontractor has long been giving 'a scheme and a price'. Why then has there been the upsurge of the method in the second part of this century for complete buildings, many quite large? Statistics are not easy to verify, but some would suggest that something like half of commercial buildings are procured this way, while it has featured enough in public sector housing for the form of contract discussed most in this book to have been developed largely in response to this demand. The consultant professions have espoused design and build contracts with varying degrees of enthusiasm, which have tended to increase as they have seen the contracts as less of a threat and as providing a worthwhile source of commissions.

A number of the reasons commonly given in an assortment of quarters, and some disputed in others, are listed below. They are given as a bridge between this chapter and the chapters that follow, where most of them reappear.

(a) Contractors are the practitioners in actual *building*. It is only reasonable that they contribute their experience of handling materials, assembling detailed parts of the work and organising the whole site operation. This should be designed into the scheme to promote 'buildability' with speed and economy.
(b) One team to build and design brings the distinct specialists into a co-operative relationship, eliminating the 'us and them' syndrome, which, it is implied, does not exist in other integrated industries. The 'contractor in the team' approach, when the contractor is appointed early or at least introduced tentatively on a fee basis, goes so far but leaves the usual gaps.
(c) The traditional pattern of independent consultants makes these gaps in communication and control inevitable, although giving the contractor responsibility for developing details within a design in which the main principles are already settled helps somewhat, provided responsibilities are clear. Achieving (a) and (b) also removes the gaps, by putting everyone under one roof, or at least one control – that of the contractor.
(d) Businesslike employers prefer to deal with a single, positive organisation covering all aspects of the building and all stages of its production.
(e) Businesslike contractors see the virtue of marketing their services in general and controlling their client market and product in the interests of efficiency and more certain profit.
(f) Design integrated with construction should produce cheaper buildings of similar standards to those achieved by the usual processes.
(g) Design integrated with and overlapping construction should produce buildings more quickly from conception to completion, particularly during the site phase.
(h) When the mainstay of the project is in standard, and possibly industrialised

construction, it makes sense to design around it, rather than to inject it unevenly into that remainder.

(i) A simple lump sum is payable without additional, and often fiddling (in the best sense) fee accounts for all the consultants.

(j) The contractor carries the risk for his design within his contract price, often covering ground conditions. The employer pays only for his own changes of mind.

It is not the aim of this book just to attack or defend the validity of these reasons. But focusing on them in a number of chapters highlights some of the confirmations or contradictions.

The questions which rise easily concern, on the one hand, matters of design. Does 'buildability' simply mean what the particular contractor is best at, and how adequately are alternatives brought forward? If the contractor has his own design team engaged from outside, as is often so, where is the benefit of integration? On the other hand, some questions are commercial. Is the employer actually receiving better value for money? If so, and if the contractor also tends to be happier about his profit margins, what *is* happening? How much or little competition is reasonable without being unfair to one of the parties? If the employer does want variations, will he pay too much (and who delineates *that*?) or will the contractor lose the integrity of his design?

It is possibly easier to see a *prima facie* case for design and build contracts having the edge on occasions when economy and programme predominate among the employer's aims discussed earlier. However, there are numbers of examples in practice where function and aesthetics have been met as well by design and build as by other means of procurement. An increasing proportion of design consultants are seeking commissions in the field, although perhaps not entirely because of such pure considerations. Their operations are considered in Chapter 3.

While intricate services may lead a project away from design and build, there are also projects where the overwhelming extent of services, especially when combined with heavy and intricate plant installations, leads quite positively to a design and build solution; the building as such has become the subsidiary element in the total concept, even if it still requires good aesthetics and quality. Again, there are projects essentially in the refurbishment field, where the work centres around specialised constructions which are routinely designed by the installers. If these specialisms occur within traditionally arranged projects they are strong candidates for nominated sub-contract, which themselves are often carried out as a sub-set of design and build (see Ch. 18).

The broad view which may be taken is that no one procedural and contractual option is best in all circumstances and an appropriate choice should be made.

Chapter 2

Principles of the design and build system

Design economy

Procedures, responsibilities and progress

Extent of contractor's design

Competition and financial control

The first chapter reviews the aims of employers and some of the contractual frameworks within which these may be met. The key principle of design and build, as against other frameworks, is that it simplifies the central contractual position to that between employer and contractor, without mediating consultants. This is achieved by allocating all design responsibility and liability in a thoroughgoing case to the contractor alone. Design and construction are contractually one 'package' supplied from contractor to employer, whatever each may arrange with their own supporting consultants acting 'behind' them. Communications across the contract divide are reduced to a single channel, which itself is a virtue. It is not concerned with further channels between each party and his consultants.

A number of general issues arise in this context which can be construed favourably or otherwise for design and build. They group under the features listed at the head of this chapter and are discussed with the aims of the employer, rather than the contractor, in mind. They introduce somewhat selectively a number of the major topics considered more technically in the rest of the book.

Design economy

This refers to the range of ways in which potentially the contractor may be able to provide value for money to the employer, either by his technical expertise or through the contractual arrangements. Whether the potential is achieved depends not only on how these ways are implemented but upon other features discussed in this chapter.

Technical integration

Much can be gained by bringing together the construction experience of the contractor with the design experience of consultants, if by so doing the result is a building which is technically more efficient in construction, and also meets other criteria. This last rider is considered later.

Efficiency

Technical efficiency may relate to the ease with which the whole building may be put together, so reducing demands on the labour force as a whole, on heavy or particularly specialised plant and on supervisory staff. These direct savings may be compounded by a trimming of the overall time for construction, so reducing the contractor's overhead commitment and increasing the rate at which he releases and is paid for his locked-up resources. This aspect of efficiency may embrace, for instance, the structural system in the earlier stages or an extensive ventilating and ceiling system later on, so that an alternative system produces greater efficiency. Alternatively it may mean a comparatively minor refinement that helps the whole rhythm of work. In multi-storey precast concrete construction, the elimination of small areas of *in situ* concrete work, with shuttering delays at each floor level (say in landings), can avoid disproportionate waiting time on site and a disturbed flow in the casting yard.

Technical efficiency may also permeate at a more detailed level in the design of subsidiary entities in the work, such as fitments and positioning and layout of services. Some of these are areas in which 'things get sorted out on site' anyway. But they also tend to be areas in which junior staff often carry out design that determines site work in a vacuum.

Producers

The crux of this argument lies in how to bring the two sets of experience – construction and design – together early enough to affect the result. If the very geometry of the building may be affected, it is not sufficient to apply a few cosmetic touches. Any system can usually be criticised, but not always changed, once it has been set up, and design is no exception. The standard answer is that feedback from past sites supplies the data for future projects. This parallels the concept of cost feedback on all types of contract to supply data for future tendering. The difference is that cost feedback is reducible to statistics for standardised pockets of work, whereas design feedback is as variable as the construction projects themselves. It may take all manner of forms from the drawn to the anecdotal, either major or minor. Repetitive forms of construction lend themselves best to such rationalisation, this being evident in housing and system building. This same problem of diversity affects cost feedback, as the statistical uniformity of fairly wide data only conceals the differences of cost due to complexity, scale, sequence and so on.

The argument then has to face the problem of transmission, if constructor and designer are separate persons. Given that many building needs are non-standard and yet have similar constructional design systems, how are critical points conveyed when they arise in one contract but are to be incorporated in another? In one building, the example of small amounts of *in situ* concrete landings in predominantly precast work can lead to a disturbance of production that is unacceptable. In another, *in situ* work around complex services may be the only sensible answer. The constructor will know the snags on the previous job, the designer knows the total concept of the next, and each of them has direct experience of only a comparatively limited range of jobs.

Three broad answers may be suggested. One is for each person, on jobs with which he is not dealing, to 'look over the shoulder' of as many persons in other situations as possible, so as to develop a 'feel' for the other side of the process – never a bad idea in any type of contract. This should help each to see the common constraints of the other and where it is most often possible to be helpful. A second answer is to have a formalised final debriefing session for each project, to learn its lessons systematically. This should be the culmination of ongoing learning (and action) during progress, and be controlled so that its purpose is constructive and its findings frank. It is in some respects a deeper and narrower version of the first answer. It should be performed in the knowledge that the persons concerned are not necessarily going to work together permanently, so that both must be explicit, thereby eliminating the need to cover for the other's weak points next time round. The third answer is to adopt a deliberately cyclical approach during design of the next project. Immediately upon briefing, one person or the other puts forward some ground rules for the design solutions. The other reacts to these and the design progressively evolves through continuous review. While decisions are grouped into phases, such as those in the RIBA Plan of Work (outlined in Ch. 5), reiteration must be constant, if it is going to be fruitful. By this means design does not become frozen at any particular stage, until the various considerations have been weighed.

The punch line of all three answers is the same: a close *working* relationship of constructors and designers, and not just some flow chart of activities. This involves quite a change in outlook, to positively aiding the other person's activities, rather than letting him 'sort out what he's paid to do'. It hardly needs an intuitive leap to recognise that this is most likely to happen when the two disciplines are part of one organisation that regularly works this way. It particularly calls for an adequate design team and not just someone who is handy on a drawing board. If the in-house facilities are not available, the next best plan is for the contractor to engage outside consultants, preferably all from one office, who have experience in design and build and who are flexible in their methods.

Products

Comment has concentrated so far on the producers, rather than on the product, which potentially can be made more economical in money and time. The obvious products to benefit are buildings into which past experience can be embodied most intensively, especially when there is a repetition of components or ideas. This gives a wide spectrum, with at one end buildings where repetition of the essential technology is absolute, among which are housing, farm buildings and warehouses, particularly those based upon factory products for the structure. The possibility of production economies are highest here, although they must be balanced against any loss of flexibility.

At the other end of the spectrum, are buildings which house complex specialised plant which entirely dictates the form of the buildings. This plant is likely to cost far more than the building into which it goes, or which is fitted around it; if the plant happens to be a 'package' scheme (all the more likely if its design is innovative), then so, logically, is the building. Laboratory and process buildings for medical purposes and facilities with unusual environmental control have led to such solutions, as have nuclear power stations, which are among the very largest of projects. Here the emphasis is not on the 'buildability' of the design, but on its integration with the dominant plant to give overall efficiency. Care is needed to see that the building brief does not degenerate into one of 'give us something to hold the plant up and keep the weather off'.

Between these extremes lie many buildings without these peculiar technical features, but with some room for integration of design and construction beneficial to both parties. What is open to exploration in all cases is the balance between effort in design and construction. This raises several issues which feature in the next section.

Commercial balance

The issues just mentioned are founded in the contractor's responsibility for a design and build project commercially and, beyond that, legally. This is the inevitable result of 'one contract and one payment'. Much of this book is given over to working out these issues in both respects. The immediate concern is commercial, although competition and control are dealt with later.

The general sequence of events is for the contractor to be briefed to design and build so much building of a certain type and quality for a particular budget. If he considers right from the beginning that these targets cannot all be achieved, it is up to him to make his views known, so that adjustments may lead to a feasible brief. Given this, the contractor should then work towards a scheme which is economical in meeting all the other targets of the employer. Design and build allows the contractor the possibility of allocating to each element of the scheme that proportion of the total cost which he considers will lead to the optimum economic result, while still adjusting this within limits if he sees the need to do so during progress. It does not lend itself to drastic design revision in order to take in major technical developments which may evolve during progress, without substantially dismantling the agreed basis of design and payment. Some projects benefit by more intensive design than others, in value for money terms – whatever 'value' is sought. This is important, in that construction accounts for 85% to 90% of the initial cost of many buildings, with design accounting for the balance, leaving aside any financing changes. Extra cost in design may repay the contractor by producing a larger saving in construction, to be allowed in his price. Design here may be expanded to include production planning, which should be linked in such decisions.

Under other contract arrangements the design effort is measured, for the employer's purse at any rate, by a fee which is commonly related to the construction effort on a percentage basis, or else is a fixed sum, even if arrived at in competition. Although the fee level varies to take some account of the scale and relative complexity of the scheme, once the consultant is appointed it is fixed. Therefore in this situation there is a consultant – there may be several – with a fee related to the budget and usually to the final cost outcome. Unless there has been a prior architectural competition for the scheme, a comparative rarity, he does not enter into competition over his *design* as such. While he will aim for a satisfactory design, or in exceptional cases for a 'monument', so long as the cost matches the budget closely enough he is not under extreme pressure over the price. How closely he has to struggle to achieve the budget depends on the employer's constraints, which may be severe. Alternatively, the consultant may have been in the happy position of helping to fix the budget! Most important, he is not in actual competition with others over the *price*, with his appointment and fee dependent on the outcome. Also, if he designs with adequate factors of safety and performance, he is still not risking his competitiveness. At worst, he faces criticism over the level of tenders and has to redesign elements on his scheme. The cost consultant too follows this pattern fairly closely.

A design and build contractor, on the other hand, while facing these pressures to some extent, faces the additional fact of competition over design, which it would appear is transmitted to his consultants as well in their appointments. This affects the quality, but is most noticeable in the price, which often has two features. One is present when he tenders under other arrangements and is his price level for a specific unit of construction, such as a square metre of a given type of wall or linear metre of a given type of cable. The other is peculiar to the design and build system in that he is in competition over which type of wall or cable he puts forward, so that his competitiveness depends on the cost per unit and also on how many units he may require in his design. This latter may be relative to the efficiency of his layout, or to the factor of safety that he assumes: with efficiency the employer may have equally as good a building with less resources used, but with a higher factor of safety he may have a 'better' building for more resources.

This relationship is illustrated in Table 2.1 (see also Ch. 6 for expanded discussion). Tenderer A has used more walling than Tenderer B but is lower in price, because he has used a cheaper walling. However, the lower price for the same cabling tips the balance on price in favour of Tenderer B over the total tender. But this is not the end of the story, for Tenderer A is providing distinctly more accommodation for his marginally higher price, and so is in the lead overall.

Conceivably it would be possible to achieve an even cheaper building if the design which is most economical in resources was built by the tenderer with the lowest unit prices. This is obviously not permissible, and might not in fact be wholly true, as the level of pricing may relate to a particular design and the use of that particular contractor's labour and plant in constructing it. In other contract arrangements, it is the price level alone that varies, as the design is constant. All this points to design competition, with its follow-through of buildability for at

Table 2.1 Effect of quantity of construction and price level on keenness of tender

	Tenderer 'A'			*Tenderer 'B'*		
Brick wall	30 m^2	£25.00	750			
Concrete wall				20 m^2	£50.00	1,000
Cable (same)	60 m	£15.00	900	60 m	£10.00	600
			£1,650			£1,600
Units of accommodation			10			8
Price per unit			£165			£200

least the contractor offering any one design, as significant factors in those cases in which design and build has a price advantage.

Total aims

At the beginning of this chapter it is suggested that the aims of the employer are being stressed. So far the discussion has been studiedly ambiguous in its references to cost, economy and so on. It has avoided the question of how costs or economies are passed on to the employer by the contractor (a process referred to later in this chapter and more extensively in other chapters). There is also another aspect of value for money concerning design, rather than price level. This relates to the total cost of the project over its life, rather than just its initial cost. These costs, such as running and maintenance already mentioned in Chapter 1, are an important element, however a building is contractually procured. They are usually a bigger consideration than initial cost. Because the contractor is competing over price as one aspect in design and build, and because this partly depends on economical design, he may well select features that bring down his tender amount. These may push up later costs for the employer. There may be a layout that is cheap but difficult to operate, or cheap materials with a short life or high maintenance costs. As a result, the apparent economy of the scheme may be false. This is as important as questions of functional suitability (with which it overlaps at the detail level) and aesthetic and related matters, although more susceptible to financial analysis.

The employer should therefore ensure that tenderers are adequately briefed on the degree of weight placed on this aspect, and he should also evaluate tenders carefully to check how they meet the brief.

Procedures, responsibilities and progress

There are several aspects in which design and build contracts are claimed to have the edge in these areas. While it may not be always so, the following are relevant because they are to an extent distinctive and all relate to the 'single bridge' relationship between employer and contractor.

Communications and the employer and contractor

Design and build contract forms provide for the employer and contractor to interact more often and more directly than other forms where, for instance, an architect and a quantity surveyor are interposed as mediating angels or the opposite. As this is the pattern in which design and build legal relationships are set, it is as well for the parties also to conduct all their earlier dealings as though it already existed. This especially affects the place of their consultants, and is discussed further in Chapter 3.

So far as the contract forms are concerned, these persons do not exist, not only in the middle role, but even as advisers to one party or the other. The nearest approach is in someone like the employer's agent, who is referred to as an optional appointment in the JCT form discussed throughout later chapters. He may come from within the employer's own organisation or be some outsider, probably a consultant. He is however an *agent* of the employer, in the normal legal sense of the word. He is not someone endowed with powers to decide issues between the parties named in the contract, binding them unless they seek arbitration to secure another verdict. Anyone like this agent is thus 'the employer's man', looking after his interests and speaking on his behalf, just as the contractor's (site) agent is always recognised as exercising these functions on behalf of the contractor. This removes some powers from an architect, say, who is acting as the employer's agent, but it invests him with others, according to the extent of delegation from the employer. Those under the JCT form are listed in Table 7.1.

Consultants to either party, in their roles as *consultants*, are moved for practical operations not considered in the contracts to peripheral positions as indicated in Table 1.1. The more technically knowledgeable employers may choose to consult them, but also do many of the front line dealings themselves, as may contractors. When they are brought up front, even if it is for all their functions, the place of consultants should be clearly defined: are they acting as agents, do they need to refer back, do communications channel through their party to the contract, can they deal directly with the other party's consultants if any, what can they decide, and so on? On the employer's side, this is particularly important over design checking, instructing changes in the scheme, extensions of time, price adjustments and payments. On the contractor's side, design is the major innovation, most other issues following a similar course to that in other contract patterns.

It is of course possible for each party to a design and build contract to go it alone without consultants. If the employer does this, either he has to be very active in responding to the contractor, who makes the running by requiring decisions, payments etc, or he may sit back and accept (perhaps by default) whatever comes to him under the contract and miss what does not. It is therefore very common for the employer to appoint a quantity surveyor, and for the contractor to recommend that he should, so there is someone who will deal with these issues. The contractor has to face up to the question of design again in deciding his needs.

One of the major points of all-out design and build advocacy is that everything is provided to the employer in one package and that he speaks to one person only. Given a simple project, a close pre-existing relationship between the parties and a solid assurance that 'goodwill is more important than disputes', then he may be able to proceed on this basis. Given even a moderately complicated project and a need for normal, prudent, commercial reassurance, he may very well decide otherwise. The process of building is both crude and complex with its own peculiar procedures, for better or for worse, so that at least *one* consultant and counsellor is valuable.

It may be countered that these are just the usual arguments for 'jobs for the boys'. However, simply rearranging where in the organisation chart people might do their work does not magically eliminate the work. Even when a function, such as design or cost control, is split so that the employer's and the contractor's people are performing part each, this may not mean duplication. Their activities may be needed to give consent to the design or check the costs put forward by the other. If the 'independent' consultants of other contract arrangements have been absorbed into one of the parties, it is not altogether unreasonable for the other to seek safeguards. This need not be 'jobs for the boys', it may just point up a weakness in the unqualified design and build argument and so give a rational balance. It is notable that most established and respected design and build contractors urge employers to appoint at least one consultant, as already noted. They no doubt view this as a benefit to *both* parties!

To have this discussion about consultants under a heading of 'Communications' which refers particularly to design and cost is no accident. For it is what is designed (and then built) with money which is exchanged between the parties, about which they talk and write throughout the project. What has been stressed so far is the place for some consultant activity with each party to help their separate interests and effective communication, without any attempt to quantify the extent, which may be assessed from Chapter 3.

But the positive argument for design and build is that much activity of a 'consultant' nature is located with the contractor under his control, and so is linked with the activity of construction. This means that he is operating on internal lines to secure economy and remove breakdowns in communication. This should benefit the employer in the contract price. So too should the fact that the contractor has no one to blame but himself when drawings etc are not forthcoming, provided his brief is complete. The employer too knows why there are no drawings: either the contractor is behind or the employer has not given prior data or consents – but there is no consultant in between to confuse potential responsibilities!

The contractor's responsibilities

The last point leads from the procedural into the legal responsibility under a design and build contract. Because the employer has one person to deal with, he has just one person to blame if the need arises. The employer must remain liable for expressing in his brief what he requires, but the contractor is responsible for satisfying these requirements once his offer to do so is accepted. While the employer should assess whether the offer is complete when stating that the contractor will meet the requirements, it is still the contractor's responsibility to achieve this, technically and otherwise (there is detailed discussion of the extent of the contractor's responsibility and liability in Ch. 5). If the windows leak, the fault is the contractor's, without it being necessary for the employer to differentiate whether it is one of design, manufacture or installation. In the unfortunate event of a really major dispute (as leaking windows actually led to the case of *Holland, Hannen and Cubitts (Northern) Ltd* v. *Welsh Health Technical Services Organisation* (1980), not a design and build contract), the single responsibility of the contractor to the employer is perhaps the greatest strength of a design and build contract, outweighing any virtues of speed and economy. It runs through a proper design in response to the brief, specifying good quality materials and workmanship and achieving them, performance in general and over the programme in particular, to meeting residual liabilities until these are extinguished at law by the due passage of time.

Whatever the strengths of having independent consultants, to whose design the contractor builds, their very independence of operation does allow uncertainty over responsibility in practice, if not in theory. This is often a question of weak communication between contractor and consultant, or consultant and consultant, where interface uncertainties arise if specifications, instructions, confirmations etc are not precise. It may also be a question of failure to operate adequate contractual machinery properly, which was what the leaks led up to in the case cited. As happened there, the employer may still have redress against the person at fault. There are alas plenty of instances in which responsibility cannot be allocated unequivocally, or the risks and costs deter the aggrieved person from seeking legal resolution. There must be many more times when the employer pays out for lesser slips between cup and lip without really knowing that he is doing so. Indeed, this is sometimes regarded by implication as a primary purpose of contingency sums, since there are grey areas between extra work required, work inadvertently left out initially and mistakes needing correction. The point here is not that a design and build contractor does not aim to cover himself for such slips between himself and any of his consultants, but that he must carry them within his contract amount, whether small or large, and cannot charge the employer extra for them.

Two caveats must be entered here. The first is that if the contractor fails to meet his obligations and also fails commercially and passes into insolvency, the value of any contractual responsibility is severely truncated. It therefore behoves the employer to check not only the general standing of any tenderer, as he always should, but in particular his design capability. This should take account of design and build work already performed and its case histories. If the contractor's design team is in-house, it is important how it is constituted and how its activity is co-ordinated with the rest of the organisation. If the team or some part of it is external, its experience in design and build working is relevant, while the question of co-ordination becomes perhaps more pressing. No such checking avoids the possibility of collapse of the contractor, but it may guard against consequential problems. Possible residual action in tort against his consultants may be available to help.

The second caveat is that the provisions of standard forms of design and build contracts do reduce the contractor's responsibilities. Thus his responsibility to supply a building meeting 'fitness for purpose' criteria may be eroded to one of designing without negligence and constructing with good quality materials and workmanship, while his performance obligations over the programme may be reduced in given circumstances. These instances are discussed in Chapters 5 and 8 over the JCT provisions, which do just this.

Time and progress

A saving in overall time from conception to completion, with this shared between reductions in design and construction periods, is often claimed for design and build projects. Some opinion puts this at two out of three projects showing a marked increase in speed, with some justification. In fact, this is usually emphasised by contractors far more than is the greater extent of their liability for faults, although this too may be given as a selling point. But then consultants do not advertise their own potential liabilities either!

There are those advocates of design and build who appear to suggest that in some magical way the system eliminates actual processes, especially pre-contractually, or at least removes frictional time loss between stages. A review of the stages in the RIBA Plan of Work, summarised in Chapter 5 and devised for traditional architect-designed schemes, will show that feasibility studies, planning applications, sketch plans, tendering and all the rest, must still logically occur somewhere in the larger pattern. Where there *is* room for a saving in programme time, is in the carrying out of some processes in parallel rather than sequentially, while internal communications tend to reduce the numbers of instructions and debates. This needs care over 'landmark' matters like planning permission, where some decisions are fundamental to all that follows and must be cleared in advance.

During design development, it may be feasible to plan the construction methods and programme fairly closely and prepare the tender financially up to a point. It may rather be that the design grows out of construction methods and cost, as part of a dynamic interaction. This is one advantage of the contractor bringing his experience into the design, while it is also something that needs watching to ensure that good design is not lost by this methodology. Factors like cost advice interact with the 'pure' design in any contract system; with design and build the opportunity is there to bring forward tendering and production in support of it, and so save some part of the time for this otherwise distinct stage. The programme saving is likely to be greatest with the more standardised schemes.

During construction, the benefits of a buildable design should show in faster building, as should those of internal lines of communication between designer and contractor to transmit basic information and to resolve difficulties. To the extent that design has not been completed pre-contractually, but overlaps with construction, there will also have been a saving. Incomplete design as such while construction is proceeding can never be a virtue: whether it will impede site progress depends on how efficiently it proceeds. In any case, the contractor carries the integrated responsibility, so that he bears the costs due to any delays of information which may cause disturbance, and also meets any damages for late completion. Here again, in arrangements with independent consultants, the employer may find himself unsure as to where the blame lies, which is one of the problems that the BPF scheme (see Ch. 13) seeks to tackle. Even if he can place blame and recover, his building is still late in either case.

Modified design

There is one somewhat separate aspect of construction: design variations or changes. In other contract arrangements, these all emanate either from the employer through his consultants or, over smaller technical matters, from the consultants themselves. Apart from their direct costs, which may go up or down, such instructions to the contractor may affect his programme by adding work or disturbing the flow of progress, so that the contractor becomes entitled to complete the whole works later. Often the incidence of variations (as they are called in such contracts) is heavy, perhaps because documents such as bills of quantities facilitate their direct evaluation, thus giving the impression that their effects on site are dealt with equally simply.

In the design and build arrangement, these changes (the term used in the JCT form, but not others, rather than variations) are treated in two ways. Those emanating from the employer, perhaps on the advice of his consultants, become instructions to the contractor and he is then paid an adjusted amount and may be entitled to a later completion date. The means by which they are valued are usually such as not to tie down precisely what the contractor receives, as is discussed in Chapter 9, but also reflect more rigorously the cost of revising the design. Those emanating from the contractor's own design team, for purely technical reasons, do not entitle him to any adjustment of money or time. If they are developments of an incomplete design then they are not strictly changes and must be produced to fulfil his contract. If they are genuine contractual changes, he must obtain the consent of the employer, who will be paying the same amount for something different.

The evidence from many design and build contracts is that the total numbers of changes and late decisions, and their effects, are far less than under other contracts. This is partly because they will be defined out of existence when they are the contractor's responsibility. But it is also the case that the contractor introduces fewer such amendments, because of their effects on his operations. The number of changes, defined as such contractually, is also less, possibly because the employer is discouraged from making them by the contractual system, and possibly because his consultants are not responsible for the design. This is enough to lead to some quiet meditation. It is also suggested that the employers who opt for design and build contracts include a higher proportion of those employers who instruct less changes anyway, but this remains conjecture. Whether less changes mean inferior buildings, or just that good decisions are made earlier, is further conjecture. They certainly mean less programmes overrunning their time.

Extent of contractor's design

An important feature of design and build is the way in which it can accommodate very varied degrees of design involvement by the contractor. If the employer or any consultant of his does no work other than prepare a brief, then clearly the contractor has complete responsibility for the design. This remains so, even if he sub-lets it to his own consultants. A comparison of the main contractual options occurs in Table 4.1, and these are set out more fully in Chapter 4.

Work by employer or his consultants

The employer should make up his mind right from the beginning whether he needs consultants to perform any of the tasks which for simplicity's sake in this book are usually attributed to him alone. Roles that they might fulfil are described in Chapter 3. Exploratory talks with likely persons and assessment of what they may provide can be very useful, and these can be extended to possible contractors whose reactions and comments may add a further dimension. The least satisfactory arrangement for employer, contractor and consultant is for the consultant to be appointed some time into the programme when decisions have already been made. The worst option is still not to appoint a consultant when affairs are out of control and fire brigade action is needed. The existence and roles of the employer's consultants should always be made known to the contractor as soon as possible, even at briefing.

Just when briefing by the employer evolves into preliminary design work is hard and pointless to define. In the following sections two situations are defined in which the employer uses a consultant: that in which the contractor takes over design of the whole project at some stage, briefing or beyond, and that in which he takes over a design of only a section of the project.

Design of the whole project

This is the complete design and build concept, in that the contractor assumes responsibility for the whole design. This he may do at any stage from briefing onwards, or he may even guide the employer in the formulation of the initial brief. If the employer has consultants, they are likely to prepare the brief and perform quite an amount of design, but this is not always the case. The more standard stages are given in the RIBA Plan of Work (see Ch. 5), but the transition can occur part way through a stage and need not be at the same point for all parts of a project. It is essential that the contractor is clear about precisely where he takes over and what the employer has already done, either tentatively or with commitment, and how far the employer is asking him to check the work and assume responsibility for it. The employer's brief and details of what he has done should be brought together as soon as possible into a comprehensive set of documents, which the JCT contract refers to as the employer's requirements.

The response or contractor's proposals may be brought in at the stage of providing a basic design or later in developing the employer's basis, or the contractor may be brought in so late that he is doing little more than provide constructional details. This is hardly design and build in concept and could lead to a confusion of responsibilities, although it is an option in the ACA contract and the only one in the BPF system discussed in Chapter 13. The JCT form with contractor's design can be stretched to fit, but its provisions over design liability are unsuitable. Other JCT forms providing for design by the architect do not place any design responsibility on the contractor; attempts to do so in the bills of quantities or specification are most likely to be invalid. The only suitable approaches here are those under the next heading.

Design of part of the project

Two broad varieties of hybrid approaches to building work are used to secure entire design of some part only of the works by the contractor, while the overall design and co-ordination are performed by the employer's consultants. These are intended for relatively larger and smaller parts of the works, although there is no absolute distinction. They are both distinct from the approach in which the contractor performs design detailing only, but for the whole project, as in the BPF system.

For the larger case of a substantial package of mixed work, it is possible to use a document such as the JCT Contractor's Designed Portion Supplement discussed in Chapter 12 to give the contract basis, when used with an ordinary JCT with quantities contract form. This latter constitutes the parent contract for all the work, with the usual arrangements therein applying over the relationships of contractor, architect and quantity surveyor. The supplement introduces the subsidiary package which the contractor designs, but over which the architect acts for the employer in the way in which the employer acts in a complete design and build scheme.

For the smaller case of an element of one type of work, it is possible to use an arrangement such as the performance specified work clauses in the JCT with quantities form discussed in Chapter 17. The work is measured in the contract bills and again the contractor designs and the architect acts for the employer.

If a JCT or a similar pattern is used in one of these ways, several comments are applicable. Such a system is necessary if any design work is to be included in a contract which is otherwise build-only, if the contractor is to be made responsible for that design. The reason for doing it at all is bound to be something unusual about the package, usually of a technically specialised nature. An alternative way of obtaining specialist design of work within a contract is by nominating a subcontractor to the contractor and making him responsible by a collateral agreement direct with the employer. This is considered in Chapter 18.

In the case of performance specified work, its extent and nature are established by the quantities provided. An essential criterion for a designed portion is that its extent and nature within the whole must otherwise be made clear. This is satisfied most directly by a separate building or other independent structure in a composite scheme. But it can be met quite adequately by foundation work, a structural frame, a services installation, an acoustic ceiling or some other discrete element, although the smaller-scale approach may be more convenient. In principle, it could be the complete superstructure of the one building in the contract, with the sub-structure and external works designed by the architect. When the proposal reaches this stage, it should be asked whether a full design and build contract would be better.

The contractor may be brought in to begin designed portion work at any stage of development from initial briefing onwards, but extra care is needed over its phasing. This is because the work has to be dovetailed into the architect's total scheme, and so is dependent on that developing sufficiently. But also the tender for the package is only part of the tender for the whole scheme. If competitive tendering is being used for the whole scheme, consideration must be given to how and when each of the tenderers is to prepare his design upon which part of his tender is based, so as to be fair in all respects. This is discussed in Chapter 12, as is the possibility of repercussions in the architect's design. For performance specified work, some preliminary specialist advice may be sought, but design proper is usually possible only when the bills of quantities go to tender.

Competition and financial control

The traditional separation of design and construction in building means that one part of the formula is fixed, as what the contractor has to produce is determined for him. The uncertainty for the employer, until the receipt of tenders, is the contract sum; thereafter, he is concerned about quality and performance, while the contractor is mindful of his costs. In a design and build tender situation, what is actually to be produced is also uncertain, since no two tenderers are offering exactly the same thing in return for the money quoted. This self-evident fact emphasises some additional considerations.

The employer's policy

A tender is simply an offer, open to acceptance so that a contract is concluded. A single tender is always open to two criticisms: it lacks other tenders to make it competitive and the most obvious means of assessing its keenness is lacking.

These points are as valid in design and build as anywhere else, but are complicated by the extra element of design. Tenderers in competition have a spur, but also have the uncertainties considered under the next heading. The employer needs to be aware that he is receiving tenders which differ over more than price, so that the perennial problem of comparing like with like is present. He can meet aspects of this half-way by stipulating as clearly as possible in his brief what his priorities or prerequisites are, and so rule out too wide a spread. At the same time, he does not want to cramp the initiative of any of those whom he has invited because of their design expertise and variety. He is likely to value the advice of an architect over the qualitative aspects of schemes and the advice of a quantity surveyor over the quantitative aspects, and the blend between them.

But while the employer may find value in encouraging competition, which is frequently used, it must be kept within bounds in fairness to tenderers who incur the considerable extra cost of design in this system. There was a time when there were reports of over forty tenders for a single scheme: this is iniquitous under *any* contractual regime, as contractors lose heavily and employers bear the costs across the range of work commissioned. The details of policy, like most issues raised in this chapter, are developed further in later chapters. In outline, the employer may consider how carefully he appraises contractors before allowing them to tender; whether he is inviting firms whose expertise, resources etc match his project; whether to invite two-stage tenders and reduce the field part way through; whether to narrow down to negotiation with a single contractor in the latter stage or even, with advice, whether to start the whole process that way.

Another issue for the employer, after controlling the means of reaching the contract sum, is that of controlling expenditure. As indicated in Chapter 1, design and build contracts are usually based upon a lump sum, which is contractually fixed, as they are special cases of a drawings and specification contract. During progress, the employer has to pay out that sum progressively for work performed and also agree any adjustments of the sum to reflect changes in the work which he requires. He may also need to agree adjustments due to changes in market costs. For all these reasons, an analysis of the contract sum is very useful, because it defines where the money is distributed and so reduces uncertainty in such calculations. The employer should decide in advance how finely divided an analysis is needed and stipulate the appropriate sections. Because tenderers produce their designs while tendering, it is impossible to provide them with bills of quantities upon which to base their tenders, because each tenderer will have different quantities. Even if quantities produced by the successful tenderer are included in the analysis, they are of limited value because of problems over errors and who bears the effect of them in the final settlement (see Ch. 9). It is therefore better to have an analysis consisting of a number of subsidiary lump sums to give a main control framework, and agree any further detail during progress. Conjecturally, this reduced initial detail may mean that changes are settled at a higher price level and that, as a result, less changes are instructed.

The contractor's dilemma

This stems partly from the employer's policy – which of course may have been adopted in response to the contractor's salesmanship! The design and build system puts the contractor in competition when computing his tender, not only over price level, but over the quantity of work in his design. This has the virtues already propounded, but does put the contractor under pressure. A two-stage tendering approach helps to relieve some of this pressure. The employer should also be explicit about the minimum information he requires.

On the one hand, he may prepare a full and detailed design as part of his tender, even if it is not called for by the employer so early, and so be able to produce as accurate a tender as possible. But if he does this, his tendering costs are high and will not be recovered if his tender is not accepted, unless a design fee is payable, which is very unusual. On the other hand, the contractor may decide to prepare only a lesser amount of design, if this is permitted, and for completion only if his tender is successful. This avoids possible wasted expense, but means that his estimating must make assumptions about what he has yet to decide over design. At the extreme, he may have to use figures almost as global as those in a cost plan, based upon analysis of other projects which he has designed and built. With an intermediate design, he may be able to use all-in prices such as per metre of *in situ* concrete column, including formwork and a likely percentage of reinforcement, or per unit of joinery fitting, which are akin to normal 'builder's quantities' prices used in drawings and specification tendering, but here still make design assumptions, albeit on a smaller scale.

How far to adopt one of these policies over design and price is a dilemma because the tender is a firm offer. In practice, the most common approach is just small-scale drawings with specification and programme, giving a good level of definition, but leaving all detailing to come. The contractor cannot include on his own initiative provisional sums for undecided work or provisional quantities for work uncertain in quantity. He can include such allowances only if the employer is undecided in his brief, or if there is a physical uncertainty such as ground conditions or the structure of a building which is being altered and if, then, it is agreed that these doubtful elements are not to be at his risk (although they often are). He must therefore cover all quantitative uncertainties as adequately as he can in his tender, while seeking to be competitive.

As far as design is not final, but the price is, the contractor has the further uncertainty of whether his post-contract design development will be accepted by the employer as satisfying what is implicit in the preliminary design and any accompanying statement of intent. If this is cast in the form of a performance specification, there should not be too much room for argument. If it is hedged about with 'adequate', 'properly' and similar dubious terms or has to be developed in essentially aesthetic respects, there is more leeway. The employer of course has a corresponding uncertainty here. He should try to say to himself: 'Is this what I would have accepted if an independent architect had designed it?' But the difference is that if he answers his own question by 'No', he knows that in that other case he could have insisted on something else, provided that he met the bill. Here the preliminary question is: 'Does the contract state that it includes such a standard?'

This question of design development is difficult to resolve. When in effect does the employer's refusal to accept that the contract is being satisfied turn into a requirement for a design change which should be paid for? Not only do standard design and build contracts wisely not attempt to give rules about it, they assume by silence that the question does not exist and so cannot lead to a dispute. This does emphasise the element of goodwill, which should be present in *all* contracts. It is all the more necessary here and is being achieved more frequently in practice.

The foregoing paragraphs assume that the employer has made clear in his pre-design and tender briefing exactly what he wants, physical unknowns aside, so that the contractor can resolve his own problems at his own discretion. If the employer is not clear over aims such as those discussed in Chapter 1, the contractor cannot resolve the priorities on which his tender should be based. He must therefore press the employer over quality and quantity, which vary directly with price, and speed of design and construction which vary in less predictable ways. Unless these tensions are dealt with, the contractor can only guess and probably offer the employer less than the best. If he is in competition, the result may depend fortuitously on who guessed most closely and not on who put in the best scheme and tender against precise guidelines.

Chapter 3

The place of consultants

The needs for consultants

The positions of consultants

Other chapters usually do not mention the existence of professional consultants, referring to only employer and contractor. In most respects consultants exercise their normal skills which do not need further discussion. This chapter highlights special features of their relationships and dealings when acting in a design and build scheme. Apart from the distinction between designers and quantity surveyors, the most fundamental one is between consultants working for the contractor and those working for the employer, and in this contractual pattern they do work for one *or* the other. A lesser distinction is between consultants directly on the staff of the employer and contractor, that is 'in-house', and those engaged from outside on a contract basis, and so properly 'consultants'. All types of designers, for present purposes, are included.

The needs for consultants

It is possible in a very simple scheme for there to be no work requiring a specialist designer, so that none is engaged. This is effectively limited to something like a small, rectangular, open plan shed with a concrete slab base on a near flat site. Even so, more work which is strictly design is involved than mere selection from a catalogue, while the availability of a consultant on an occasional basis may still be valuable. For anything more elaborate, one or more consultants on a continuous basis is prudent, and becomes a necessity at some scale.

The contractor's needs

The contractor will be aware of his needs at a lower threshold of scale, as he is producing the whole scheme and is the more likely to engage a full range of consultants. As the project threshold is raised and the frequency of projects increases, so he must also face the question of when it will become economic to have his own department. The economics are obviously related more to continuity of work for such a department, otherwise staff will be underemployed. This will more than negate the savings that might flow from the avoidance of external fee accounts, and no doubt sometimes do! Perhaps less obviously, there are the cost benefits of internal communications, of which the saving in people's time, even though significant, may be the least. More important may be the benefits of feedback etc discussed in Chapter 2, which lead to more economic and better designs which win more contracts.

A difficult level for the contractor is that of organising enough work on a continuous basis for only a restricted department. If he cannot employ a sufficient range of staff to perform all aspects, and staff of sufficient calibre at that, he may be ill-advised to undertake design, particularly internally. This is a new function for any contractor entering design and build work and it must not be underestimated over the scale required, its complexities or potential liabilities (see Ch. 5). It is best for the contractor to keep cost control with his own quantity surveyor, whom he is likely to employ for other purposes anyway, and who can also co-ordinate separate consultants in project management terms.

If the scale and intricacy of design are too much for this approach the contractor could consider an architect on his staff to brief and control external design consultants. The contractor needs *some* liaison whenever he engages outside consultants, so that he can forge the best link between construction and design. It is advisable to secure experienced staff for this function when operations demand it, and not rely on someone already in his organisation in another capacity 'filling in', however good he is at his own job. When he has such staff, he should continue using them in this way until he reaches the threshold for a full department, of which they may become the nucleus. If anything he is better to pass the threshold by a safe margin before making the change. An internal department is the logical result of enough growth and continuing prospects, but not before. Even when it is achieved, highly specialised design may call for outside consultations on occasions.

The employer's needs

The employer has quite distinct needs from those of the contractor who, in performing the design, also seeks to control his own costs. In essence, the employer wants the right building at the right price and in the right time. As price and time are each a single, fixed, contract figure, it might appear that only the building design can vary in detail, with the employer just keeping his eyes open to get what he wants. Therefore, why duplicate costs by having another set of consultants beyond those of the contractor?

Design

In fact, the employer's first problem may be that of knowing what in outline is going to be the right building; that is, establishing the brief for the designer. This relates to function, appearance and other matters raised in Chapter 1, and to the earliest stages isolated by the RIBA Plan of Work (summarised in Ch. 5). Here an architect is accustomed to drawing out the employer's ideas and establishing a pattern with priorities. If the employer intends to have a design consultant in due course, it may be better to bring him in at once to provide a brief, in full or in outline.

The contractor's own architect will also be able to do this and a good design and build contractor can provide a comprehensive service right from the beginning. Limitations arise when the contractor is not providing the expertise required or frequently when, with or without expertise, he leads the employer towards those solutions which suit his own building methods. If there is to be competition among contractors, multiple briefing and comparison of their suggestions can become cumbersome, although more ideas will be generated. The employer's consultant can filter out completely unwanted options and save abortive work, but should stimulate any attractive alternatives. It is significant that many reputable design and build contractors advise employers to engage at least one consultant of their own (not always a designer), whether there is competition or not. They obviously value having someone with whom they can share a common language and who can forestall problems.

With the brief established, any consultants of the employer can watch over the employer's interests without the counter-responsibility of looking after those of the contractor as well. This will extend to ensuring that design development does not go off course, assessing the detailed offer or offers, and ensuring that the contract standard is met during construction. Again, there is the value of a common language over the work involved in examining constructional detail and checking technical performance.

All of this is not a duplication of the contractor's design work, but an extension to include the design elements of liaison and quality control that the employer's consultants perform under other contract arrangements and which are still needed here. If it is valid to argue that financial savings accrue from bringing design so much closer to construction, it is also likely to be valid that moving design within the contractor's control leads to some direct extra expenditure by the employer to give a degree of security in a commercial transaction. In the end, it is the trade-off between the two that is important to the employer.

Price

It is true that price is more simple than design, in that the tender amount which becomes the contract sum is one figure, but before and after this transition the story is different. At the briefing stage, the budget must be fixed at a level that relates properly to the proposed scheme. It must then be kept under review through design development right up to acceptance of the tender, so that the budget and design are adjusted as desirable to keep in step. This is an area in which the contractor must obviously be active and can provide much data; but this statement begs the question of how the employer is to be sure of a sound deal as value for money. Competition is the obvious spur to keen tendering, but in this type of contract it rests on the provision of differing solutions as well as differing amounts. At each stage and whether there is competition or not, the employer stands to gain from the activities of his own quantity surveyor, who can set up a framework for the budget and then evaluate, compare and negotiate, so that finance and design may be related.

When the contract is in being, these same activities are needed with any design changes, less frequent though these are, and various other matters that lead to adjustment of the contract sum. As competition has now ended and the employer is committed, the need is the more obvious even when affairs run smoothly. If, however, there are disputes over disturbance and lapses by the employer, again less frequent, the need becomes pressing. While the employer may be at a disadvantage in dealing with design on his own, at least much of it is visible and tangible: costs do not readily share this attribute. In fact, if the employer has to settle for one consultant only, it is not unknown for him to choose a quantity surveyor for these reasons.

Time

Time also is a single figure, if viewed simply as the period from conception to completion. But this ignores the time needed for each stage between, with the inevitable delays which may occur for various reasons. A good design and build contractor will prepare a realistic programme, which he is best able to do in these circumstances, although the employer's consultants can well check it. The employer, however, may need protecting against himself if he is offered an attractively short programme, which in fact allows him inadequate time to assess properly what he is being offered and, perhaps, less obviously, to make later decisions as they arise especially during construction. If there is to be competition, the need for some regulating time-scale that is fair to all pre-contractually, but not excessive, is plain. Even a minimal contribution from an architect or quantity surveyor may be invaluable here, while an intricate, phased scheme may warrant far more. If the programme goes wrong post-contractually, there is the question of who bears the extra cost of the work and the effects of late completion. The parties are by then committed and these are aspects where matters often spill out of the tidy compartments shown in the programme and the contract sum analysis to give rise to disproportionately large extra amounts of time and money. An adviser who has been acting for the employer over the corresponding routine issues is able to help here.

Conditions

While, in principle at least, the employer can withdraw at any time up to entering into the contract, thereafter he is bound by its terms over the foregoing issues of design, money and time, and others too. It is therefore important for him to have a contract that is fair concerning his interests and someone who is able to advise him or act as his agent over its interpretation and operation. It is best for him to enter into a standard form with terms which are at least well known, if not clear or infallible. His consultants are able to advise on which form to use. If the contractor puts forward his own form, they can check it for difficult or biased provisions and advise whether to use it with any amendment or whether to reject it and seek another.

A review of the employer's needs shows several areas in which consultants may be of assistance, to put it mildly. These involve distinct applications of skill which relate to, but do not merely duplicate, what the contractor is doing. The contractor has absorbed design and any liabilities for it and so has removed the role of *independently acting* consultants. The employer still needs technical guidance and perhaps the protection afforded by those who act for him alone in this situation. When a consultant is engaged by the employer, the distinction must always be maintained between what he does if acting as agent for the employer under the contract and what else he may do as adviser to the employer outside the contract, over such issues as the suitability of the contractor's design in aesthetics, function and value for money.

The general stance of the consultants in this contract pattern follows that of the respective parties.

The positions of consultants

While exercising their normal skills, consultants are responsible to those who engage them in new roles, to iron out any potential ambiguities introduced, not always explicitly.

Working for the contractor

Architects, engineers and other designers

When any of these persons are on the staff of the contractor they are clearly his servants and their working procedures are affected accordingly. But when they are engaged as external consultants for a particular design and build project by the contractor, their position is again changed. (The question of liability is discussed in Ch. 5.) The contractor commissions all design and is responsible to the employer for the integrated design and construction, while the consultants have no contractual responsibility to the employer over design (though there remains a possible liability in tort, also considered in Ch. 5). They also have strong responsibility to the contractor over competitiveness and cost: the former meaning obtaining the contract, and the latter making a profit.

Procedurally, an external or staff consultant is similarly affected. He takes over any prior design embodied in the employer's brief and deals with it as part of the total brief given him by the contractor. He is thus not dealing with a 'lay' client, but with one who is knowledgeable about the technicalities of building and who is most likely to lay down a whole range of essential or desirable features about the project. These may well include major aspects of constructional form or materials, impinging upon economy of layout, ease of building and the various matters introduced in Chapter 2. These features may extend into areas such as sub-contracting, which are not the designer's primary concern under other contract patterns.

The consultant must maintain adequate standards of design, as any responsible contractor will require him to do, while accepting that the scale of priorities may be different. Ease of production in some instances may be more important than a detailed refinement in layout or aesthetics. This may be true in other types of contract, but there the designer can usually carry the day. Here he is conditioned by feedback from other projects and, in the end, by the wishes of his client, the contractor. He is also in no position to tell the contractor that he must wait during construction for design information: there is no equivalent of the architect's power to instruct postponement of work as in other contracts!

How this relationship is worked out will be affected by several considerations. If the consultant is within the contractor's direct organisation, he must establish for himself and his department a definite level of authority over decision-making within the organisation, in relation to those dealing with policy, planning, estimating, construction and so on. This may come out by formal decision or by informal negotiation whichever prevails in any power structure, but usually by both. If the consultant is external, his authority needs formal definition every time, although if he is performing many commissions a regular pattern should emerge. This relates not only to what he *may* do by way of requiring certain design solutions, but also to what he *must* do under his contract of engagement and also what he *need not* do.

A further consideration is how the employer's brief, when given to the contractor, is transmitted to the designer. The order of events may be: employer to contractor and then contractor to designer, with the contractor 'translating' the brief on the way. If the designer is on the contractor's staff, he is likely to 'be' the contractor for briefing purposes and so short-circuit matters, subject to observing organisational ground rules. But even if he is an external consultant, it is possible for him to receive the employer's brief direct. Whether he should receive it *alone* is very much open to question, in view of the integrative aspects of the design and build process. The designer might prefer direct briefing alone, because this is what he is used to, but he has to learn to think another way round. The contractor has to consider his own aims and how a consultant can best further these. The contractor must learn to appreciate what a designer, especially an architect, is trying to provide which is not directly economic and how much freedom he needs to achieve it. The best product, which will help sell the next one, will be a blend of contributions – preferably not just a compromise by a committee!

This last consideration is in turn affected by the stage at which the contractor takes over the design function, whoever actually performs it. There is no fixity over this in design and build, as is considered in Chapter 5. The earlier the contractor's designer comes in, the nearer he may be in being briefed to operating as he would in other circumstances. The later he comes in, the more he will be constrained by earlier decisions, whether made by the employer, his consultants or the contractor. He should check carefully what may or may not be amended out of what has been done and whose acceptance of changes is needed. If he is to incorporate design as provided to him or base his developed work upon it, he should never put himself in the position of accepting responsibility for what he cannot verify. In particular, he should be watchful over any production of working details and portions of design which he performs based upon concepts of function worked out by the employer's previous team, to ensure that he does not assume responsibility by default for what he has not originated.

The message of the preceding paragraphs is clear for both contractor and consultants: clarify and agree relationships and procedures in this adaptable approach to building. Clarity should dispel woolliness, but not produce woodenness. For without clarity, integration may beget confusion.

This discussion has concentrated on design itself. But design turns into construction, and with it goes quality control and supervision. It is not unknown for external consultants to be given duties of site inspection by the contractor, with powers to reject poor work. This must be expressed in the appointment, as otherwise the consultant is acting on site literally as a consultant, if he is there at all.

Lastly, there is the matter of design changes during construction. The contractor should establish a clear routine over who may introduce these, designer or site staff, and whose consent is needed within his organisation, quite apart from that of the employer outside. On this type of contract, as on any other, it is as undesirable for changes to be injected part way along the production process, say by the site agent; although exception may be made over details below some threshold, if it can be defined. The designer in turn, while correcting mistakes, should be wary of changing his mind. This is never popular with a contractor, but at least, in other contracts, he is paid for the consequences. In this version, he suffers both the annoyance and the loss. He is in a better position than most employers to tell what has happened, *and* to recover from the designer at fault.

Transfer of consultants by novation

A possible and quite common situation is that the employer engages his own consultant (most likely an architect, but possibly more than one) to prepare the scheme design as presented for a tendering response within the employer's requirements (see the discussion in Ch. 5) and then requires the contractor, as a condition of obtaining the contract, to take over and use the same consultant for the rest of the design work. This requires the substitution of a contract with distinct terms between the consultant and the contractor, and it is often achieved by novation of the original contract from the employer to the contractor and so by agreement between all three parties. It is important that the first contract contains the provisions to be included in the second, to avoid any hitch in making the transfer or even it being rejected. Not uncommonly, the agreement will contain terms by which the consultant warrants to the contractor the adequacy of the design which he has prepared for the employer.

Such a transfer allows a potentially smoother flow of design work from one stage to the other, as the design does not have to be picked up and developed from another's concept.

A disadvantage of such a transfer is that it can be made to only one contractor, so that it is not available during tendering when there is competition between several tenderers. Further, there is an awkward overlap of need for the consultant's services from completion of the employer's requirements up to entry into the building contract. The architect, or whoever is involved, is working at preparing the contractor's proposals and yet the employer needs his advice over those proposals when submitted. The architect will find himself with a conflict of loyalties unless he disengages from the employer, contractually and in procedural terms. The employer must go to another architect if he needs guidance at this stage, and so loses continuity in this way.

During the post-contract phase, the same distinction must be observed. While the employer has nominated someone who was 'his' architect, he must see that he now cannot call upon the architect's services and certainly does not now have a consultant as a form of Trojan horse within the contractor's defences. On the other hand, the contractor has acquired a consultant who is there to serve his interests and who takes instructions from him alone over the design and related matters. It is conceivable that in points of relative detail the consultant may be required by the contractor to modify his original design from what the employer had approved. This he must do, even though he should warn the contractor, with it being a matter between contractor and employer to negotiate the solution to any perceived conflict. The account of consultant duties under the next heading underlines the overall position.

Working for the employer

Designers

The salient distinction here is that these consultants are not only paid by the employer, as is normal, but are paid to look after his interests alone. They do not therefore have to be continually deciding issues, as for instance architects have to do when issuing the various certificates called for under the JCT forms. Their conduct must remain professional, but they owe a duty to the employer only; the contractor already has the option of his own consultants to advise him.

The result is the employer's consultants are unable to dictate to the contractor: they can say nothing to him at all unless authorised by the employer, and unless the contractor is prepared to listen. The most that, for example, the JCT form with contractor's design does is to allow the employer to appoint an agent who acts for and virtually 'as' the employer in whatever functions the employer delegates to him, but without any right of independent action under the contract. For this reason, other chapters make few references to consultants.

The employer may choose to use his consultants in any way he wishes behind the scenes, but in relation to the contractor their erstwhile functions are severely truncated. Thus it may happen that they are not brought in until some while through design development, or even during construction. Late introduction does suggest that problems have cropped up and that 'fire brigade' action is needed. Consultants approached in these circumstances should define their terms of engagement and responsibilities with due care.

It is to be hoped that consultants will be involved in drawing the brief out of the employer, formulating it for presentation (perhaps in stages) to the contractor and then in assessing the contractor's scheme prepared in response (again perhaps in stages). But they cannot dictate to the contractor what he is to produce, in design or building, without danger of losing the advantages of the design and build approach. If they do then, even more on their own doorsteps, there is the further danger of assuming expressly or by implication an actionable measure of responsibility for the contractor's design and its failings. Not only can they not *dictate*, they should not *approve* (as is said of the employer repeatedly in later chapters), but merely note the contractor's proposals as apparently satisfying the brief. They are advising the employer and not running the project design for him.

The same philosophy applies to the construction phase. If the employer has nominated them, his consultants may inspect the works during progress over quality, and instructions may be issued over defective work or materials. It is an activity which the contractor should welcome, as many do, because it can prevent bad work passing into the works as handed over and surfacing as defects much later and much more expensively. The standard is conformity with the contract. If the contract documents are precise over what is to be provided then normal procedures are followed. But if design was incomplete when the contract was formalised, then the contractor has some measure of discretion within whatever was actually said about what he is to provide. To this extent, consultants have to stand by and advise the employer over the reasonableness of the provision. They cannot insist on a particular solution, unless they can persuade the contractor that no valid alternative exists.

Consultants should ensure that their design is completed up to whatever stage is provided for in their appointment by the time that they become *functus officio* with the appointment of the contractor as designer. They have no opportunity to pick up loose ends during subsequent stages (an activity which unfortunately some consider routine in traditional working) and certainly not to correct faults. If these lie in the basic concept, they may lead to very expensive actions from a dissatisfied employer. However, it is important that consultants keep watch over the development of the design at any stage in which they are involved as employer's agent (see Ch. 7) to ensure that design is not deviating from the concept which they prepared for embodiment in the employer's requirements. This can be particularly important if the contractor or other designer goes out of business or dies, when the employer's natural recourse may then be against his supervising consultant for not ensuring that all was well. Between the two jagged rocks of incomplete design and design compromised later with disputed responsibility lies the channel to the clear seas of a satisfying project all round and, not insignificantly, a clean record of professional indemnity.

These considerations call for a more detached attitude and, it might be thought, a more negative one. But in fact the employer's design advisers need to see themselves as free to stimulate ideas and not just turning down what they view

as half-baked initiatives. Their position requires a more dipl
mixture of firmness and humility – and which of us is any th

Quantity surveyors

Far less need be written about the broad position of quantit
design and build arrangements although, like designers, their ties
may be picked out under the guise of 'employer' and 'contractor' in other chapters.

This brevity stems partly from the fact that *design* responsibility has been shifted. Quantity surveyors acting for either party in relation to design are likely to find themselves giving cost advice. The employer's quantity surveyor may well be required to establish the budget and a cost plan which will be of value to both parties in establishing a controlling framework and avoiding wasted effort. As always, his role is advisory, but also he must not relieve the contractor's designers of their responsibility for designing to the budget. The contractor's quantity surveyor must work most closely with his designers to ensure that the design develops properly. He almost inevitably will be in-house and so able to filter data fed back from other projects. If the contractor is using external design consultants, his quantity surveyor will need to exercise extra cost co-ordination on that front also, as the consultants will not have the same familiarity with the contractor's organisation. Both quantity surveyors are likely to follow cost activities through into preparing the contract documentation, which is usually provided partly by each party.

Also there is little reference to the quantity surveyor in building contracts anyway. His function in the post-contract phase is not to take the initiative, but to react to other's activity or lack of it, by calculating interim payments and final adjustments to the contract sum. In other contracts he is assumed to stand between the parties for these purposes, so that he can exercise his impartial function. In practice he usually has to maintain the balance by first negotiating with his unmentioned opposite number in the contractor's organisation, who has only the contractor's interests as his responsibility, besides his own professional approach. The quantity surveyor named in the contract then has to try to stand back and weigh matters in the balances by being fair to both the contractor and to the (silent) employer. Under a design and build contract, any quantity surveyor to the employer acts solely for his client and so is relieved of the necessity of performing the balancing act. His practical activities remain essentially the same.

Quantity surveyors who find themselves called in late in projects for 'fire brigade' action need to review their terms of engagement at least as much as designers. They are quite used to this sort of role.

ter 4

Standard contracts allowing contractor or sub-contractor design

JCT standard form of building contract

JCT contractor's designed portion supplement

Some principles relating to the two foregoing JCT documents

Clauses within the JCT standard form

Domestic sub-contract

BPF/ACA system and agreement

Conditions of contracts GC/Works/1 design and build

ICE design and construct contract

JCT nominated sub-contract forms

This chapter is descriptive of the contract forms available, which are discussed or referred to in Parts 2 and 3 of the volume. Table 4.1 sets out a skeletal comparison of the choices available. Each of the forms directly indicated or implied there is mentioned in this chapter, in varying detail, and each heading is referenced to the table and to later chapters. The table order shows the relationship in principle

Table 4.1 Typical range of design delegation to contractor and sub-contractor, under various contracts

Contract	*Envelope*	*Services*	*Floors*
(a) JCT standard	Architect	Nominated SC	Nominated SC
(b) JCT designed portion	Architect	Nominated SC	Contractor/Domestic SC
(c) JCT performance specified work	Architect	Nominated SC	Contractor/Domestic SC
(d) ACA first alternative	Architect	Named SC	Named SC
(e) ACA second alternative, BPF/ACA	Architect pre-contract .... Contractor/Named SC/Domestic SC post-contract		
(f) JCT with design (inc. with BPF supplement)	Contractor/Domestic SCs		
(g) GC/Works/1 design and build	Contractor/Domestic SCs		
(h) ICE design and construct	Contractor/Domestic SCs		

Note: The first four examples assume that maximum delegation has been selected. Under (b), the 'floors' are 'the portion'. The last four examples are rigid for the employer and the contractor has flexibility, except if a 'Named SC' is used under (e).

between contracts, while in the main the order in the rest of this chapter and book reflects the way in which contents are compared. Part 2 deals with the sequential development of practice aspects and relates the clauses of the JCT with design contract to that sequence.

JCT Standard Form of Building Contract, with Contractor's Design, 1981 Edition

(See Table 4.1, item (f) and Chs 5 to 10, 14 and 17)
This document is issued by the Joint Contracts Tribunal (JCT) and published by RIBA Publications Ltd. References to the JCT form, contract clauses, conditions etc are to this document, throughout this book, unless it is otherwise stated or obvious. The JCT is not in any way a tribunal, but consists of representatives of some of the bodies covering clients, contractors, sub-contractors and consultants. Inevitably, their interests are not always identical, but they are committed to negotiating documents (the full range is legion), for general use and on which they are unanimous. This means consensus usually by compromise, thus achieving middle ground on many issues in the eyes of many, although by no means all.

This contract is a thoroughgoing design and build form providing for all design to be performed by the contractor, without any alternative. This is subject to anything given in the employer's requirements (explained hereafter) as briefing data, which may be quite extensive. In keeping with this, it excludes the concept of nominating sub-contractors or suppliers, and indeed makes no explicit arrangements about the employer's requirements limiting the choice of sub-contractors or suppliers in any other way. Other JCT forms mentioned in the next paragraph deal with, or at least overtly recognise, each of these options.

The form has been produced as a variant of those other JCT forms which are usually known as the Standard Form of Building Contract (see Table 4.1, item (a)). These exist in editions as 'with quantities', 'with approximate quantities' and 'without quantities', each available in private client and local authority variants, making a total of six forms. The present form has a similar philosophy and provisions, consistent with its purpose. As a result, there is more in this volume on clauses in the present contract with those aspects which constitute the differences. Some comment has been given for rounded treatment on other key features, but readers who are not already familiar with these in detail and need to be, should consult a fuller treatment, such as the present author's *Building Contracts: A Practical Guide*.

The contract is issued in one edition only to suit both private and local authority employers, thus differing from the main series standard forms. Its format is similar, containing in one binding what are strictly five contractual documents.

(a) Articles of agreement, which evidence the contract. There are a number of insertions of names, amounts etc to be made in obvious gaps.

(b) Conditions, which contain standard clauses as the bulk of the document. There are several alternative clauses and clauses which apply only in particular cases, all indicated by footnotes. Unwanted clauses *must* be deleted for the conditions to read correctly. A code of practice referred to in the conditions follows.
(c) Supplementary provisions (issued February 1988), which are unexplained but have been included in response to the British Property Federation's system for building procurement. They read as additions to the main conditions and modify or qualify these without need for amendment. They are discussed in Chapter 14.
(d) Appendices 1 to 3 which allow for insertions and choices of alternatives. These give substance to various references in the conditions, which themselves do not need amending. In most cases the appendix items *must* be dealt with to give any effect in the contract.
(e) Supplemental provisions, which constitute the VAT agreement. These are referred to in the conditions, but not in the articles, and they are not separately evidenced. They always apply, despite their semi-detached relationship, and govern value added tax as a matter semi-detached from the contract.

There is no list stated as of contract documents in the articles or, as in other JCT forms, in clauses 1 and 2.1. The point is considered in Chapter 5 but the list of documents in clause 2.1 may be taken as the contract set:

(a) Articles of agreement.
(b) Conditions.
(c) Appendices 1 to 3.
(d) Employer's requirements.
(e) Contractor's proposals, with the contract sum analysis annexed.

This list does not include the supplementary and supplemental [*sic*] provisions in the main stream, by definition, although the related sub-contract accommodates value added tax *within* its clauses. The last two sets of documents listed are those specially produced, even if partly from the parties' own standard material, for the project which contain all the technical data. Because they came into existence earlier, they may contain information on how the other documents are to be completed, but this should not be a substitute for completing the other documents. This, and the risks in making amendments to standard documents and the relationship between the several documents, are taken in Chapter 5.

The JCT document is supported by two JCT practice notes issued separately from the contract and from one another:

(a) Practice Note CD/1A, which contains notes on how this contract is intended to be used and on the main steps to be taken in bringing the documents into completeness and a coherent relationship.

(b) Practice Note CD/1B, which contains an outline commentary on the form of contract and guidance on the contract sum analysis and formula fluctuations.

Reference to these practice notes is made in this book. They are not part of the contract and do not therefore affect its legal interpretation, although conceivably the courts might bear them in mind if in some circumstances they bore upon the clear intentions of the parties. An example might be to decide what was contained in each of the several parts of an unclear contract sum analysis. Otherwise their status is, with respect, no greater at law than that of this book! Further, they are not so worded that they could be incorporated in whole or part into, say, the employer's requirements or contractor's proposals by direct reference, much as happens to the Standard Method of Measurement under some JCT contracts. Any principles that they contain could become contractual only if redrafted as express rules etc within the requirements or proposals.

JCT Contractor's Designed Portion Supplement (1981) to the Standard Form of Building Contract with Quantities, 1980 Edition

(See Table 4.1, item (b) and Ch. 12)

This document comes from the same source as that just outlined. Because of its nature, it does not need extensive separate treatment and most points relevant to it may be obtained by reading discussion of the contractor's design form. Its purpose is to introduce into an otherwise traditional architect-designed scheme a portion which is designed by the contractor, as a sort of pearl within the oyster. Thus the parent editions of the JCT form are those 'with quantities', either for private or local authorities use, to which the present document is no more than a supplement grafting in the distinctive design and build features, so that they apply for a portion of the works only.

The supplement as printed is not suitable for amending other JCT forms, although it would be feasible to produce a document that was, subject to copyright conditions.

The document in one slim binding consists of:

(a) A revised first part of the articles of agreement (up to and including article 1), to be *substituted* for the corresponding part in the parent contract.
(b) A schedule of modifications to a further article, to numerous clauses and the appendix, to be read *in addition* to the parent contract, introducing insertions, deletions or substitutions into the latter, without the need for direct alterations.

In this case, clause 2.1.2 as substituted does list the contract documents as:

(a) Contract drawings, which will cover the rest of the works and at least indicate the boundaries of the contractor's designed portion.
(b) Contract bills, which may include the amount for the contractor's designed portion without detail, although alternatively the amount may be taken to a separate summary with a total of the contract bills.
(c) The employer's requirements, which may refer to the contract drawings and bills for further detail.
(d) Contractor's proposals, related directly to the employer's requirements alone.
(e) Analysis of the portion of the contract sum relating to the contractor's designed portion.
(f) Articles of agreement.
(g) Conditions.
(h) Appendix.
(i) Supplementary appendix, mentioned separately as the rest of the contractor's designed portion is not mentioned.

Some principles relating to the two foregoing JCT documents

An important spur to the publication of both forms for the first time in 1981 was the need of public sector housing for a balanced design and build arrangement. There is, however, nothing about either form that either restricts it to or favours such use. Both forms relate to the 'modified entire contract' concept, as outlined in Chapter 1. Both also allocate complete design responsibility to the contractor, whatever the employer's requirements that may have been supplied to him, for the work which they cover contractually, comprising either the whole or part of the works.

In doing this, both forms leave completely open the extent of the employer's requirements, so that Practice Note CD/1A can say that these 'may be little more than a description of accommodation required, or may be anything up to a full "Scheme Design"'. Whatever work the employer has done must, however, be included in the employer's requirements: perhaps obtaining partial or full planning permission and design or specification that constrains the contractor's work. The contractor's proposals must complement the employer's requirements by bringing all design etc up to completion or indicating sufficiently which direction it is to take to meet the requirements. They must therefore be a reasonable extension of the requirements and may, within the limits set, modify their detail.

Another aspect common to both forms is the provision by the contractor of an analysis of the lump sum for the scheme in his proposals, the form and level of detail of this analysis being as stipulated in the employer's requirements. This is to be used for specified purposes only, relating to payments to the contractor. There are further common aspects in the absence of nominated persons and the inclusion of minimal stipulations over the contractor's power to use domestic sub-contractors. Anything more might compromise his design function, although in the case of the contractor's designed portion he may well face nomination and greater domestic control in the rest of the works.

Several differences flow from the use of one form for the whole works and the other only for a part. The fundamental one is that when the supplement is used the employer has an architect responsible for designing and specifying the rest of the works, and not an agent interacting with the contractor on a more limited front, as under the full design and build form. As a result, the supplement is concerned with allocating responsibility for design between architect and contractor, and also liability for any related default in performance leading to delay and other consequences, in a way that is unnecessary in the other form.

When design is split, it is not only necessary to allocate responsibility within the parts, but also to include power over co-ordinating the parts. This power must rest with the architect, who is cast in a more impartial role than the employer's agent, but there is obvious room for tensions when the power is exercised (as by definition it must be) post-contractually. This is an area where contract terms can hardly resolve the difficulties, and matters must depend on mutual goodwill – or legal proceedings! The way forward, happily or otherwise, lies in a modification to one or both designs to secure one form of harmony. The architect needs to consider whether to modify his design to take in a reasonable development in the contractor's design, or whether the contractor may reasonably be expected to modify his design while keeping within the proposals prepared in the light of the employer's requirements. In the former case the employer pays a different amount; in the latter he does not. The point is discussed further in Chapter 12.

While the supplement allocates responsibility, it does not give any guide as to what sort of 'portion' it is to govern. Its use could therefore range from a complete building to an element, as the practice note suggests. The main criterion is distinctness of the physical entity and its design performance, which form part of the discussion in Chapter 12. An alternative approach for smaller elements more closely bound into the architect's overall design is provided by the clauses for performance specified work mentioned under the next heading.

Clauses within the JCT Standard Form of Building Contract with Quantities, 1980 Edition

(See Table 4.1, item (a) and Ch. 17)

Within the private and local authority with-quantities editions, but not within the approximate quantities and without quantities editions, there are clauses to deal with 'performance specified work'. This is a means of obtaining direct contractor's design of elements of work under what is otherwise a build-only contract. The work may be measured or covered by provisional sums. Otherwise, design may be obtained by using the contractor's designed portion supplement as an addition to the contracts or, more indirectly, by nominated sub-contracts as the contracts regularly provide (see in this chapter and in Chs 17 and 18).

Domestic Sub-Contract DOM/2 Articles of Agreement, 1981 Edition

(See Table 4.1, items (b) and (f) and Ch. 7)
This document is approved by three bodies, representing various main contractors and sub-contractors, and published by one of them, the Building Employers' Confederation. Although it is not a JCT document, it is produced as a sub-contract available *optionally* for use *only* with the JCT with contractor's design form. It is not referred to in the main contract. It covers situations in which the sub-contractor is or is not required to be responsible for designing the work he performs, in part or whole. It is not suitable for a design-only case, where the usual document for appointment of the consultant concerned should be used. In principle it is equally suitable for sub-contracting work under a JCT contractor's designed portion supplement, where nothing else is available, but in practice needs a large number of minor verbal changes to give it precision.

The document is in one edition and in one binding with these sections:

(a) Articles of agreement, which fulfil the usual purposes and which are the substantive part of the document, as its title indicates.
(b) Appendix in several parts, which gives particular details of the sub-contract and main contract, similar to what is provided for other domestic and nominated sub-contracts related to JCT main contracts.
(c) Schedule to sub-contract, which consists of numerous amendments to the conditions of Domestic Sub-Contract DOM/1 after the manner of those in the designed portion supplement.
(d) Schedule of DOM/2 Sub-Contract Conditions, which is simply a list of clause titles.

Sub-contract DOM/1 mentioned in (c) is another optional sub-contract for use with the other JCT main contracts, in the absence of both design by the sub-contractor and nomination by the architect. It is produced in two bindings, one of which consists of the sub-contract conditions and is of comparable bulk to the JCT design and build form. The effect of the schedule to sub-contract is to incorporate the DOM/1 conditions as the conditions of the present sub-contract (which otherwise has none, only the list of titles in (d)) and to amend them to suit the DOM/2 situation, over design matters, and to relate to a different main contract. As a result, it is not strictly necessary to include a copy of the DOM/1 conditions in the sub-contract documentation, although one should be to hand so that its wording may be consulted (Sub-contracts DOM/1 and DOM/2 are discussed in the present author's *Building Contracts: A Practical Guide*).

No practice note exists for this sub-contract, which is referred to in Chapter 5. It is the last of the documents relating to the JCT design and build concept proper.

BPF/ACA System and Agreement, 1983/1984

(See Table 4.1, items (d) and (e) and Ch. 13)
The documents mentioned here and referred to selectively in Chapter 13 are issued by two bodies, the British Property Federation and the Association of Consultant Architects, who hold the respective copyrights. They have been prepared unilaterally by the parent bodies, with only limited consultation with members, rather than representatives of other bodies, and might be construed *contra proferentem* by the courts. The major documents are:

(a) Manual of the BPF System, for Building Design and Construction, 1983. This describes the system and sets out its rationale and procedures in some detail. It is thus partly in the nature of a practice note, following the JCT terminology. It declares itself to be a 'purely advisory and not a legal document' and so not a contract document, nor interpretative of any. It does, however, assume the existence of consultants' agreement, a building agreement and sub-contracts, if needed.
(b) ACA Form of Building Agreement, British Property Federation Edition, 1984. This has a similar structure to that of JCT contracts, although not so defined, incorporating articles, conditions and appendices. There are a number of alternative clauses within the conditions and several spaces to complete to give periods and percentages. The related sub-contract is not mandatory, but obviously its use is an advantage.

As might be inferred, there is also a separate ACA Form of Building Agreement, Second Edition 1984, which is for use other than with the BPF system. It is so similar in the relevant effects that, for convenience of treatment, it is not discussed separately within the scope of Chapter 13. Its main difference is that it allows a contract without any contractor's design, as an alternative to an amount of such design.

The BPF/ACA agreement contains only the provision for some contractor's design, as the BPF system requires. This is not a full design and build arrangement for the contractor to design part of the works as a self-contained portion, as with the JCT supplement. It is rather a provision for him to take over the partly developed design of the whole works *after* tendering and entering into the contract, and then to be entirely responsible for completing it.

General Conditions of Contract for Building and Civil Engineering GC/Works/1, Edition 3, Single Stage Design and Build Version

(See Table 4.1, item (g) and Ch. 15)
This document was issued by the former PSA Projects Ltd acting as consultants to the Department of the Environment. It is a unilateral document, although

consultation with interested parties throughout the industry took place, and therefore open to being construed *contra proferentem*. As prepared it is suitable for use by government departments in procuring construction work, while it is also intended for adaptation for use by other clients seeking its distinctive features. As its title indicates and as is explained in the non-contractual introduction bound as part of the document, it is for use when 'design' takes place in one stage completed before 'build' follows. In all, the one binding consists of:

(a) Contents.
(b) A non-contractual introduction, which explains the rationale of the contract and its broad operation. This includes the single-stage approach and states that considerable amendment would be needed for a two-stage approach. It is also stated that the aim is for price and time certainty when a project is adequately defined, relatively simple and not to be subject to much change during progress.
(c) The conditions themselves.
(d) Two indexes: one general and, significantly, one covering the time limits set in the conditions.
(e) Three model forms: an abstract of particulars (equivalent to JCT articles), a tender and a set of essential insurance requirements.

The contents also refer to 'Supplementary Conditions – General Note', but this is not included thereafter.

The contract is defined in the conditions as:

(a) The conditions of contract.
(b) The abstract of particulars.
(c) The authority's requirements ('authority' being equivalent to the JCT employer).
(d) The contractor's proposals.
(e) The pricing document (equivalent to the JCT contract sum analysis).
(f) The tender, evidently meaning simply the form of tender of which a model is given, but which refers to the other documents.
(g) The authority's written acceptance.

The contract is formed by submission of the tender (the form for which does not drop out of the picture as in the JCT pattern) and its acceptance by letter without the use of any articles of agreement. It thus behoves the contractor to be aware of how quickly a formal contract may come into being.

ICE Design and Construct Conditions of Contract

(See Table 4.1, item (h) and Ch. 16)

These conditions are issued by the Institution of Civil Engineers, the Association of Consulting Engineers and the Federation of Civil Engineering Contractors and the result is therefore a consensus document. They are an adaptation of the mainstream civil engineering contract which has been in use for many years, with little amendment by comparison with what has happened to the JCT and GC/Works/1 contracts. The contract is intended for the design and construction of works, without any particular limitation on what proportion of design rests with the contractor or how many stages of design or tendering there may be. It gives a lump sum contract subject to variations, whereas the parent ICE contract gives a measure and value contract, without a contract sum, to take account of the inherent uncertainties of civil engineering works, such as tunnels and work below water level as so largely performed in the ground and in other fairly hostile environments.

The document is in one binding and consists of:

(a) Contents.
(b) Index to the conditions.
(c) The conditions themselves.
(d) Form of tender.
(e) Appendix to the last, equivalent to the JCT appendix of contract details.
(f) Form of agreement, equivalent to the JCT articles.
(g) Form of performance bond, which is optional.

Labour tax fluctuations are covered in the conditions, but further fluctuations need to be dealt with by supplementary conditions.

The contract documents are given by the form of agreement as:

(a) The form of agreement itself.
(b) The conditions of contract.
(c) The employer's requirements.
(d) The contractor's submission (equivalent to the JCT proposals) and the written acceptance thereof.
(e) A space is given for 'the following documents'.

The last allows for any assorted documents, although it is to be hoped that most will fall into either the requirements or submission. It is as well to list there the unmentioned parts of the present printed contract document, as their significance might be overlooked by one party when about to enter into the contract or there could be later doubt over what was included. In particular, the form of tender is the only place where the contract sum is stated, as it is missing from the agreement. The appendices of the form are referred to in the form itself.

JCT Nominated Sub-Contract Forms, 1980 Editions

(See Table 4.1, items (a) and (b) and Ch. 18)

These documents are issued by the JCT for use with nominated sub-contracts related to JCT standard forms with or without quantities. They cannot be used within either of the JCT design and build arrangements outlined earlier in this chapter, although they can be used in the other part of a contract which includes a contractor's designed portion. They are used to give a contractual relationship arising out of the nomination process, which is peculiar to the construction industry. This process is discussed in Chapter 18 in relation to the question of obtaining design performed outside the main contract, but paid for under it as part of a nominated sub-contract sum. Here, it need only be said that, of the JCT forms concerned, neither main nor sub-contract makes any provision for design from the contractor's side of the arrangement. Out of the forms listed here, only the agreement deals with design.

Five documents are provided and are obligatory to operate the scheme:

(a) Tender NSC/T: Standard Form of Nominated Sub-Contract Tender (Parts 1 to 3).
(b) Agreement NSC/W: Standard Form of Employer/Nominated Sub-Contractor Agreement.
(c) Nomination NSC/N: Standard Form of Nomination Instruction for a Sub-Contractor.
(d) Agreement NSC/A: Articles of Nominated Sub-Contract Agreement.
(e) Conditions NSC/C: Nominated Sub-Contract conditions.

It must be said here, and it is repeated and enlarged upon in Chapter 18, that whenever there is design, the agreement is essential to protect the employer's interests. It also protects the interests of employer and sub-contractor over several other crucial issues mentioned, but not taken in detail, in that chapter. (For treatment of these matters, again see *Building Contracts: A Practical Guide*.)

Part 2

Practice and the JCT contract

Chapter 5

Briefing, design and documents

How the employer and contractor join forces is not a prime concern of this book. It may be through previous contracts, recommendation, desperation, the glossy advertisement or even the proverbial golf club. Desirably, it will involve an assessment of the contractor's inherent suitability and current capability to perform the work required. Much of this is common to assessment for other contractual forms, but should especially relate to his experience in design and build projects and the structure of his organisation with an emphasis on how specifically it is constituted to perform design and build.

Critical here will be what calibre of design expertise is in-house or otherwise how it is to be obtained and utilised. From earlier discussion, it will be clear that the latter is much more than just 'we will engage such and such an eminent firm; see how good their buildings look'. It will include how the two or more organisations will interact in terms of policy and control, where the latter in particular will reside in one or another according to what function is under control, such as overall design, buildability, economics or production. If the contractor is comparatively new to design and build, he may have production departments still wedded to operating in isolation from designers or even from one another. How the contractor and his designers, external or internal, propose to relate to the employer should emerge. This sort of investigation needs some deep interviewing, distinctly more than can be accommodated in the recesses of the golf club. Visits to the contractor's offices and works and to projects completed and in progress provide flesh to clothe the skeletons of discussion. References and consultation with other clients, particularly those contacted other than by references, can reveal much.

Many contractors are quite adept at marketing in the design and build field, this meaning much more than attractive brochures. This is partly because not all contractors operate in it, so that those who do perceive the need to make their expertise known. It is also because there are proportionately more first-time clients seeking design and build or at least open to it, than exist in other areas of demand for construction with the exception of speculative housing. Unless they are told, they may not know where to start or even what they might get if they try. Preliminary consultations over a project in such instances should be viewed by the employer as part of the contractor's marketing process, as they will usually be particularly cordial (in any sense of that term) and the professed abilities need as objective evaluation as is available. It is here that an employer's consultant is valuable. The operation can continue by way of a multi-stage tendering procedure, as is considered in Chapter 6.

Often matters may slide from an exploration of what might be, into the first stages of briefing without one or both parties being aware that it has happened. This may well be fine, but it is always an advantage to keep suitable notes of meetings, so that decisions are not made and then lost to sight. If only one party recalls them, there is the possibility of that party working on assumptions which the other did not intend.

The financial aspects of striking the bargain and of tendering and any competition are taken in Chapter 6. The present concern centres mostly around design by a single contractor: how the employer communicates what he wants, how the contractor responds and what responsibilities are involved. In concept design precedes tendering and construction, although a design and build arrangement permits an overlap.

Briefing the contractor

While formally briefing may be seen as a one-way process from employer to contractor and divided into discrete and successive stages, in practice it is usually more of a dialogue between the parties, moving overall in one direction, but tracking reiteratively over the same ground from time to time.

This can be a desirable state of affairs, as it allows the employer to clarify his thinking and probably to modify his ideas. He may have little knowledge of how to prepare a brief, so that a question and answer process initiated by the contractor best sees him through. He may have his own consultants right from the beginning so that he first briefs them, perhaps again by question and answer, and they in turn brief the contractor in a more obviously structured way.

This is also desirable because the contractor's response in design will develop over time and lead to reconsideration of what the employer has asked or the contractor has prepared. While particular major decisions must be frozen at given dates to avoid endless reconsiderations, both parties should keep open to feasible changes and not sink into the 'my mind is made up, don't confuse me with facts' syndrome. Early decisions may seem tightly logical, but later may be seen on review to be leading to illogical or at least sub-optimal results. This is not peculiar to design and build projects, but needs watching when the whole briefing process is more variable between projects than with the traditional pattern, as may be seen from a comparison of the next two sections. The point is also accentuated when there is competition for a design and build contract (which this chapter is generally ignoring), because there may be several design solutions with ideas arising in one which can legitimately influence the employer in responding to another, without any question of infringement of copyright. Eventually, though, tenderers should be in no doubt over what they are doing and over what documents, and in what detail, are to accompany their tenders.

The result of the whole process of briefing and design may be embodied formally in a set of documents, principally drawings, schedules and specifications, prepared solely by the contractor. If the employer's briefing has been at all intensive and especially if it has stretched into drawings and other technical data, it will usually save laborious re-presentation of this material if the contractor incorporates it as given into his complete scheme with any qualifications to it that have arisen. This will be useful when the contractor completes design after the award of the contract, so that parts of the brief are still 'live'. The JCT design and build system recognises this by the concept of employer's requirements which form contract documents as well as the contractor's proposals.

Briefing does have an inherent framework within which flexibility and reiteration occur and the next two sections look at this from distinct angles.

Principles in briefing

The extensiveness and intensiveness of briefing may both vary by almost any amount. The employer may say little more than 'I require a building in which to do my own thing', which thing he then defines. When the 'thing' is unusual or the project has the potential to be individualistic, this may be almost as much brief as the designer may initially wish for, so that his imagination may have free play. Given a more routine project with predetermined characteristics, the designer may welcome some measure of precise guidance, subject to some leeway for change. JCT Practice Note CD/1A recognises this wide band of options: there 'may be little more than a description of accommodation required, or may be anything up to a full "Scheme Design"' given in the employer's requirements.

As a basic principle, it is desirable to tie the contractor down as little as possible in layout or technically in order that he may use his best methods. But the aims which he is to meet must be defined absolutely clearly, as must the fixed facts of the situation. Frequently the practical tendency is towards the full scheme. A middle path for defining what is essential is for the employer's advisers to develop matters up to a concept at RIBA Stage C or D (see under 'Patterns in briefing' below). This cuts down on excessive work by all tenderers and points them clearly ahead, but it does so by embodying the main features in a fairly well frozen scheme. Another way of keeping excess costs down is to introduce two stages of design development into the competition and eliminate most tenderers after the first stage, as is considered in Chapter 6. These variants take account of the fact that costs have to be paid for somewhere, even though it may not be in the price for the current project – which may still bear those of other tenders.

There are therefore several main groups of information which the employer must provide. More details are given under the next heading, but the groups may be defined here to isolate their nature.

The first group is that of the employer's priorities over the development: these relate to aims, such as described in Chapter 1, and to their relative importance. He should assess the criteria on which he is prepared to modify his aims, if he cannot achieve them all as originally formulated, in the light of the expected trade-off. This gives the contractor a picture of the margins within which he is to determine the balance of his scheme. If there is competition, more schemes will fit the employer's requirements better if everyone has just the same clear statements to use as a basis. Key elements are the quantity of accommodation (desirably in square metres and broken down into types), the quality of building and its life span, the budget (initial and total) and the time to hand-over of the project.

The second group comprises questions of fact about the site, which admit of little amendment. Physically it has a position and boundaries, and there may be the results of surface surveys and test borings, with details of services etc on or adjoining the site. Operationally it may be subject to imposed restrictions. The employer may have his own stipulations about use or misuse of the site during construction, especially if he has existing buildings or is using the site in some way. He, and so the contractor, may be constrained by his neighbours, while there may be legal elements such as easements or building lines to observe.

A third group also concerns the site, but in terms of how the project is to relate to it. It is a further set of factors affected by the interaction of the first two groups and includes the permanent use of the site in terms of the disposition of buildings, accesses, car parking and so on. With these go any requirements about the shape and height of buildings, and relative details such as elevational appearance and special stipulations about working conditions and safety, including any particular materials and specification which may sometimes be needed. But it is worth repeating that, as far as possible, the performance needs should be given and the designer left to respond to them.

A final group consists of matters which are out of the design field, but which may occasionally affect the design or budget, even at the early global stage. These include organisational matters, such as phasing of the project and limitations on temporary use of the site and on working hours and methods. They certainly should be raised as part of the briefing process: a ban on dusty operations may well call for an emphasis on off-site fabrication with design to suit.

These groups are concerned successively with more and more detailed matters, so that the settlement of elements in earlier groups is desirable before moving to later ones. It may also be seen that alternative possibilities at later stages may lead to modification of earlier decisions, on the recycling principle already mentioned.

There may be items of work or other obligations for which the employer wishes to defer a decision over what is involved or just does not know initially. These may be covered within the contract by provisional sums or quantities, provided these are included in the employer's requirements only. Their principles are set out in Chapter 9, while the ICE contract taken in Chapter 16 allows for them specifically.

In all these matters the employer must either be accurate or state clearly and with disclaimers the limits of what is given, to avoid any presumption leading to later claims of misrepresentation.

Patterns in briefing

Underlying whatever sequence of briefing may evolve, there are a number of discernible stages in the process which may be isolated, even though they may overlap or each may not happen at the same time for all parts of a project. The elements just considered tend to fall into these various stages. The best known categorisation is that given in the RIBA Plan of Work. This is used as a basis below for the stages to show how the briefing of a *contractor* as distinct from a *consultant* affects matters. The main relevant parts of each stage are summarised with comment over when the contractor may be introduced, and which is the main variable feature.

Stages A and B – Inception and feasibility

The contractor must be involved in both stages or in neither, as they go so closely together. If the employer has no consultant, full involvement of the contractor is inevitable. If he has a consultant, an involvement of the contractor with a clear division of initial responsibilities is optional but desirable: in any case the contractor will need to go over some of the ground. When several tenders are being sought, the position becomes more difficult, so that any consultant of the employer may work best on his own at this stage, unless radically different solutions are expected from the various tenderers.

The former stage covers discussing with the employer the general nature of the project physically, its time-scale and overall finances, and forming a plan of action. As the contractor is going to tender a single amount to the employer for the whole project, it is not necessary to agree between them now what work is to be the contractor's responsibility for payment purposes. He will take on whatever the employer's consultants are not going to do and allow in his tender. Each party must deal separately and solely with his own consultants. As soon as he has enough information, the contractor should place his main proposals about who is to perform his design before the employer, to ensure that there is no misunderstanding affecting later actions, especially when post-contractual approval of subletting is required.

Also in stage A, information about the site is to be obtained to clear matters of ownership, especially of accesses, restrictions, both physical and legal, on its use and development, existing services, buildings, etc. A rudimentary survey and appraisal of the site will reveal what information is needed.

Stage B is a first response to what has so far been essentially information from the employer, even if elicited by a questionnaire. It involves a study independently of the employer of his requirements, leading to a dialogue with him over alternative modes of design and construction, with an 'order of cost' indication of the financial effects for the capital and whole life costs of the project. If the employer has started by treating with quite a number of contractors, he may find it possible to eliminate several, even at this early stage and so avoid wasted effort all round. Any contractor who finds that the types of proposals that he will be able to develop are clearly not going to be of interest to this employer, should sensibly eliminate himself. Such a forthright policy should encourage further relevant business, rather than the reverse.

If the contractor is acting alone here, he will need to check whether the employer is aware of the labyrinth of planning, building and other statutory requirements that has to be traversed. This is an area in which at least one of the parties must be knowledgeable or well advised, almost more over the procedural than the technical aspects. Delays and wasted work happen only too often.

Stages C and D – Outline proposals and scheme design

If not already involved, the contractor may be brought in to deal with the former of these stages. This covers moving from a simple assessment of feasibility to setting up a control framework that takes in the analysis of the employer's requirements and relates them to outline proposals on how to meet them, say in terms of the general utilisation of the site with numbers and relative massing of buildings. This the employer should approve with any amendments needed, as giving the ground rules, along with at least 'an approximation of the construction cost', as stated in the RIBA Plan. This expression ignores running and similar post-construction costs, which it has been suggested should be in view even at stage B, unless a present discounted value of all costs has already been used. Either way, it is better to keep all costs in view at all stages, to avoid any danger of the design being unduly affected by initial costs alone, while recognising that the employer may be restricted by what he can fund immediately.

If the employer's own consultants have been dealing with stages A and B, even if in consultation with the contractor, it gives continuity for them to take what they have distilled and do the work of stage C. The argument for the contractor in any case to perform stage D is however strong, even though the JCT practice note allows the employer's consultant to do more. This stage *is* a scheme design developed from the outline proposals to show not only the size and character of the project, but also the spatial arrangements, such as layouts and sections, materials and appearance. When the accommodation is closely determined by or for the employer, it is necessary to give the constraints of sizes or arrangements to the contractor but, if not, to leave him to interpret the brief that has already evolved. This accords with the philosophy of allowing him freedom to develop his own optimum response. This even more is true when selecting materials etc subject to any absolute dictates over critical performance or appearance in specified places.

Other elements in stage D are a cost estimate and dates for starting and finishing the construction work, which obviously determine the overall pre-contract programme as well. These elements are as important in most cases to the employer as the character of what he is to receive. The cost affects the design considerably, while the time affects the cost noticeably but also the design, either through the cost of speed or by the use of construction techniques that directly affect construction time. In view of these considerations, the employer should regularly consult with whomever is designing to keep all elements in balance. He should be left in no doubt as to the extent of overrun in cost or time that will flow from changes of mind above a given threshhold, so that he can judge whether or not to introduce them.

One of the major arguments for design and build, commented upon several times in earlier chapters, is the potential for the contractor to influence the practicality of the project as a buildable entity. If therefore he is not used significantly in stage D, this opportunity cannot be available. Hereafter, he is just adding the finishing touches, unless he revises earlier decisions so leading to additional, if by then desirable, work. This is close to the position under the BPF arrangement considered in Chapter 13 and holds similar risks of divided design responsibility, which should be avoided.

It is at the end of stage D that planning permission should be sought, whatever preliminary discussions have taken place. It is a matter over which the effects are dealt with specifically in the JCT clauses taken later in this chapter. In practice employers frequently take their activities through to obtaining planning permission, because of their close concern in detail with the outcome. It may also save time to clear the ground at this stage, rather than to have to go back when several tenderers have developed designs, and simply because it may be quicker for the employer's architect to deal with it. This has to be balanced against loss of flexibility for tenderers in producing competitive design, and on this score is something to be avoided if practicable.

When the employer's consultants are acting through but not beyond these stages, they should ensure that they discharge all their duties completely. These are stages when the most fundamental aspects of the project are determined, so that potential liability for later inadequacies runs high. The contractor, in turn, needs to ensure that he is not implicitly assuming responsibilities for any such basic flaws in what he takes over. Even without flaws, a scheme which is incomplete to the point of ambiguity can lead to later doubt over responsibility. A dissatisfied employer could well act against both consultant and contractor, if in doubt as to where the responsibility might lie. The problem for the consultant shifts, but does not disappear, if there has been a novation of his services to the contractor (see Ch. 3).

Stages E and F – Detail design and production information

These two stages *must* be performed by the contractor, otherwise he will not design at all! First they encompass settling all points of design, including those of specialists, and then producing all drawings, schedules and specifications to enable the project to be built. All this information is also made available to the employer, at least under the JCT provisions.

It is in stage E that the sequence of the Plan of Work, as set for other lump sum contracts (with or without quantities), is acknowledged to be open to change by the JCT provisions. This is because the scheme which is embodied in the tender at stage H, and so forms the basis of the contract, may leave parts of the detailing and other information to be completed post-contractually. (To which the informed reader may retort that this is simply what happens in practice in other contracts anyway, and not in just those based on approximate quantities and prime cost.) Stages E and F may thus run on to overlap with stage K. The implications for the contractor of advancing or retarding design before tendering have been looked at in Chapter 2. They have implications for the employer as well, and he should balance the two sets together when stipulating in the employer's requirements the amount of detail that is to form part of the tender. Before accepting the tender, he has the opportunity from his strong bargaining position and with no tender accepted to check any detail that forms part of it and to raise objections if he so wishes. In accepting the tender, he accepts statements of intent about the outstanding parts of the design upon which the *fixed* contract sum is based, even though the design is incomplete. Thereafter it is a question of interpretation of these statements as the design is developed. The JCT form, for instance, ignores the possibility of actual disagreement.

During these stages, the contractor continues to clear statutory approvals, again overlapping with construction. The RIBA Plan mentions co-ordination of specialists' design and obtaining quotations for their work: these are domestic matters for the contractor to handle.

The other mention is of cost checking. It is necessary for the contractor to perform this for his own purposes to see that he is keeping on target. It is desirable to keep the employer informed before the tender is submitted on how affairs are proceeding. Any aberrations may be due to changes which he requires and he has a right to know, so that he may change his mind again if he so wishes. They may be due to differences that the contractor is finding during design development. It is usually as well to meet the financial problems part way, rather than when the tender is seen by the employer.

Stages G, H and J – Bills of quantities, tender action and project planning

Little comment is needed here. The employer's requirements, contractor's proposals and contract sum analysis, under the JCT terminology, take the place of the bills of quantities and other tendering documents and are sufficiently discussed elsewhere. The employer's requirements are all that the contractor formally receives, since he prepares the rest. There is nothing special about their issue, although it may well be progressive. The procedure may be affected in detail by whether there is competition up to this final point or not, and if so whether it is single or two stage (see Ch. 6).

Project planning in RIBA thinking is concerned with entering into the contract and ensuring that all the main participants know what the others are doing. Here, the latter is mainly a case of defining the roles of any consultants to either party and setting up lines of communication.

Stages K and L – Operations on site and completion
These pass beyond briefing and design as operations, even though they may be partly contemporary under design and build. They present no differences in the immediate context, other than that the information flows from the contractor to the employer. In other chapters, the complete responsibility of the contractor for his design and its construction is a recurring theme.

The JCT contract: an aside

This is the first chapter in which provisions of a specific contract form, the JCT Contract with Contractor's Design 1981, are considered in substance rather than there being simply incidental references to them. In general in Part 2, this is done without reference to its supplementary, BPF related provisions considered in Chapter 14. Two broad points may be emphasised here about the contract, applying also to other contracts.

(a) There is no reference in the contract clauses to the way in which the contract position has been achieved. Such questions as briefing, competition, pre-contract discussions and revisions, financial calculations, concessions, balancing of advantages and gentlemen's agreements are not mentioned. The only evidence recognised by the contract is that of documents as embodied into the contract. Even here, reference is made to only a few categories of documents to give them defined statuses and functions, so that any stray letters, memoranda of meetings and so forth containing contractually relevant material must be brought within these categories. This may be done by attaching them to and referring to them in mainline documents, or by writing their substance into these documents. Otherwise they may be taken into account in legal proceedings only in extremely rare instances when the contract cannot be construed reasonably even though if not as intended, without them.
(b) With the formalising of the contract, matters between the parties reach a position in which many things are fixed and are no longer options open to negotiation or even withdrawal. Changes in many matters remain possible, but on the conditions binding both parties – often at a price!

These two points are basic contractual principles, but are particularly important in a contract that may well have been preceded by extensive interaction between the parties. This interaction forms much of the subject matter of this chapter and the next, significantly the longest two in this volume. In other forms of building contract the employer's advisers take full responsibility for piloting him through many of these activities, and continue to act beyond. This arrangement may not avoid changes and extra payments, but it often anaesthetises the patient!

Design responsibility: the JCT contract

'Responsibility' is used here to denote the range of contractual obligations of the contractor, at whatever stage he is introduced, while 'liability' is used under the next heading to denote his position over any failure to carry out his responsibilities, either under the contract or more widely. 'Design' is also taken to embrace obtaining any outstanding planning permission, specifying and such interlocking activities. This section summarises the main sweep of responsibility in relation to the JCT form (which is referred to here, without reference to its supplementary provisions, as noted above), but which is broadly common to contracts of this type, although curtailed in contracts under which the employer's advisers carry design to some fairly advanced stage before the contractor develops it to finality (see Ch. 13 for instance).

Pre-contract activities

The starting point for the contractor lies before the contract proper. This is in the interpretation of the 'Employer's Requirements', first mentioned in the recitals of the contract, which are viewed as a single set of documents made available to the contractor by implication at one time, although in practice they may evolve in discussion with him, especially when there is negotiation. They should express the employer's needs, any ideas (as a minimum) that he has about meeting them and any restraints that he is placing on the contractor's response. They are likely to include the project budget or some other financial expression. There may be provisional sums (discussed in Ch. 6) for expenditure which the employer wishes to reserve for later decision and any amount of data in support of the broad statement of requirements.

The contractor's response, again viewed as a single set of documents put forward at one time and mentioned first in the recitals, is in the 'Contractor's Proposals' which set out his intentions for meeting the employer's requirements. The form and level of detail of the contractor's proposals should be as stipulated in the employer's requirements, so that they take over where the other documents leave off and give a coherent scheme. They probably will not give a *completed* scheme, in that further design development and production information is to be performed post-contractually and quite possibly after construction has begun. They must not, however, contain further provisional sums introduced by the contractor. He must give a tender which covers all his design uncertainties on a fixed basis, including satisfying any statutory requirements. The exceptions in clause 6.3 are meeting changes in statutory requirements in general and conforming with planning permissions and approvals in particular, all happening after the date of tender, although the employer's requirements may remove planning matters from the list of exceptions. Beyond these, he has no contingency sum to be spent at the employer's expense.

With his proposals, the contractor has to provide the contract sum analysis in whatever form may be prescribed in the employer's requirements. This may be

useful in assessing the tender, but is specifically for post-contract use. It should reflect the current design and cost distribution, but it will stand, however the contractor may amend his design and distribution on his own initiative during its development.

What the employer receives in the contractor's tender and, subject to any agreed amendments, what becomes part of the contract is a scheme for which the contractor is completely responsible, unless the employer has stated that he will remain responsible for any preliminary work. This is so by virtue of the contractor taking over and checking what the employer has provided, and incorporating and developing it with his own design. Exceptions may be matters like planning negotiations carried through more commonly by the employer, when there is to be competition and which the contractor must take as given. Site and soil surveys and sectors of prior design for which the calculations etc are not made available to the contractor, must again be taken as given. This sort of information should be avoided wherever possible, for the contractor should normally have been given information with a qualification that he is responsible for checking it (and so must have the means to do so) whatever use he may make of it. Otherwise there is the danger of split responsibility, especially over design, as may happen under the BPF system (see Ch. 12).

While the employer should examine and assess the tender, if needs be commenting on it and negotiating any amendments, he should not formally approve its detailed contents. If he does, he may well be undermining his own position, if a question of design liability arises. He should certainly point out any divergences from his requirements that are not acceptable to him, because the contractor's proposals override the employer's requirements once they become contract documents. But he should avoid even an implication that by so doing he approves the rest of the content. He should go no further than the terminology of the third recital of the articles, that he is satisfied 'the Contractor's Proposals and the Contract Sum Analysis [taken as an overall picture] ... appear to meet' his requirements.

This position over 'appearances' is difficult to maintain in some respects. It obviously covers design which has yet to be completed. It should also cover any completed detail design with a heavy concentration of, or high sophistication in, its technical content, such as the foundations, structural frame and services installations. It may also apply to the specification of materials like finishes where their fitness in use is not apparent to the employer, say in acoustic treatment. On the other hand, he may be faced with the contractor's choice of a critical floor finish and be aware from previous experience that it is likely to prove satisfactory. Even so, he should give his consent, but not his approval, qualified as 'without relieving you of your obligations'. These matters are taken further in Chapter 6, but see also legal cases later in this chapter.

Where it becomes impossible for the employer to shrug off complicity in the contractor's design is over matters such as layout, room heights and general massing of buildings. In such questions of space, function and impact he must be as aware under this type of contract, as under any other, of what he is being offered, and so must be held to accept it knowingly. Equally, although not with such precision, he should be aware of the broad standards of quality of what he is being offered. If not, it becomes impossible to adjudge later whether the contractor has met his responsibility of achieving those standards in such detailed matters as acoustic treatment and floor finish.

There are thus areas where the employer may not and others where he may plead ignorance if a post-contract dispute arises. In the former, the broad issues, it is imperative that the employer should check that his requirements are being fully met in the contractor's proposals and that there is no divergence, over which the later documents would be definitive. But even in the latter, the narrow issues, the employer is held by the framework of the contractor's proposals, so that he can dispute only departures from the framework and not developments within it. He can instruct a change over any issue, but the resulting financial adjustment may go against him.

Post-contract activities

When once the contract has been entered into, responsibilities are not open to negotiation in the same way as before. The contract sets out a number of standard rules to be applied over the non-standard earlier design work, for its interpretation, development and change as necessary.

The contractor's responsibility is to complete his design and turn it into production information, which is the thrust of clause 2.1 over carrying out the works and 'for that purpose' completing 'the design'. As a by-product of this, by clause 5.3 he is to provide the employer with copies of what he 'prepares or uses for the purpose of the Works'. No purpose is assigned to this transfer of information, whereas clause 5.5 requires the contractor to pass over information for specific purposes at completion. Clause 5.3 is a mirror-image of the clause in other contracts, where clearly the architect must give the *contractor* drawings and so on, so that he may construct the works. Here the provision gives the employer an opportunity to check that the contractor is not deviating from the true path of design development, or that he is not lowering standards. Not only is this not said, but curiously the whole contract is silent on this issue and on any procedure about related disagreement, other than in the optional supplementary provisions. There can of course be arbitration, but not during progress on this issue. The nearest approach is in clause 8.1, where the contractor is debarred from *substituting* any materials and workmanship without the employer's consent, once the originals have been specified. This does not cover the initial specifying, even post-contractually, and its relation to design is likely to be peripheral. Contracts from other sources covered in Part 3 are more positive over design examination and bear comparison alongside the present supplementary provisions.

Presumably the silence is to avoid putting the employer in the position of actively approving the contractor's design, which he should not do, as much at this stage as previously. When the contractor sends something purely for information, perhaps a bar-bending schedule, at most the employer should indicate that he has noted it. In the case of something in the nature of a facility, such as an enquiry counter or a kitchen extract hood, the contractor may well expect or even ask for something more informative. Here the employer could consent to the general arrangement as suitable, while still not approving the constructional details and performance (again see Ch. 6).

The contractor is also responsible for clearing all outstanding statutory approvals etc under clause 6, including planning matters. The value of the employer taking these as far as possible as is consistent with not restricting tenderers' creativity has already been mentioned, particularly in terms of avoiding delays. There is no requirement for the contractor to report back to the employer over the outcome, although most matters will arrive via the clause 5.3 copies. If some minor difference arises out of conformity with an individual building regulation, he may well not be concerned. If there is a change in the statutory requirements themselves or if the granting of planning permission amends the contractor's proposals, neither of which could the contractor have taken into account when tendering, a change arises 'as if it were an instruction of the employer under clause 12.2'. The resulting adjustment of the contract sum protects the contractor or the employer but, either way, the employer still has no sure early means of knowing what is happening. The financial consequences might be large and, particularly in the case of a planning matter, the employer might also wish to instruct some other change of his own to give him a better solution. He should protect his interests by stipulating a warning procedure in his requirements.

It is the general responsibility of the contractor to rectify discrepancies in the scheme and to absorb the effects within the contract sum. This applies over statutory requirements to the extent just indicated and applies under clause 2.4.2 to discrepancies which are entirely generated by the contractor within his proposals; however, the employer who has accepted the proposals can choose which option to have if he so wishes. Not all discrepancies can be resolved quite so simply. If the discrepancy lies in the employer's requirements, clause 2.4.1 turns the tables: the contractor's proposals prevail in their interpretation. This costs neither party anything, except perhaps the employer a measure of disappointment or the institution of a change. An error underlying what was included in the employer's requirements would lead to an instruction changing these requirements or, more costly, correcting the results. A divergence over the definition of the site boundary is inevitably corrected at the employer's expense under clause 2.3.1. In all of these cases, the party discovering the clash is obliged to notify the other. Nothing is said about non-discovery, late discovery or suppression of knowledge.

Further, nothing is explicitly stated about divergence between the employer's requirements and the contractor's proposals, when each set of documents is harmonious in itself. The employer has accepted the latter, as recorded in the recitals, as 'appearing to meet' his requirements. They will prevail, if they differ unambiguously from the employer's requirements, when these are also unambiguous. When the contractor's proposals are being assessed, it is therefore a case of *caveat employer*!

Apart from changes forced on the parties in the ways described, there may be changes desired by one or the other. If the contractor wishes to change his design as developed, the conditions once more are silent. Only clause 8.1 already mentioned referring to materials and workmanship is available. The employer should be ready to make concessions here: it would be unreasonable not to allow some discretion to the contractor and it would be nonsense to seek to hold him to a design which has been found to be faulty. Concession or not, the employer is always entitled to receive at least as good a deal as the contract provides, unless he is prepared to receive less and pay less. If, however, by some change the contractor is able to provide the employer with a solution which has the same user value to him but costs less, it is reasonable for the contractor to be paid the same amount. What is actually his second solution could well have been his first, and no question would have arisen. It is the spirit of a design and build contract that the contractor is at risk on both elements, and risk runs both ways for him. It can be argued that he tenders at a level allowing for one or two cost refinements in his own favour, without which the incentive to efficient design might be stifled. It is a rare employer who wishes to pay more in the reverse situation.

JCT Practice Note CD/1A expresses the pious hope (which is achieved quite often in reality) that the employer's requirements will have been prepared in the expectation of few subsequent changes, the term used formally in the conditions in place of 'variations', as in other forms. Changes are defined as instructions of the employer requiring the contractor himself to change his own design and do other things outside this immediate survey. Such changes are dealt with in clause 12, along with the instructions over expending provisional sums included in the employer's requirements. They both lead to financial adjustment.

The contractor is entirely responsible for responding in design terms to an instruction in these areas. If it is a change, he has the right of reasonable objection to accepting it at all which may protect him against several undue effects, but particularly against a disturbance of the integrity of his design for which he could not fairly be responsible. He has a right to immediate arbitration here, but might instead be prepared in a marginal case to accept the instruction, subject to an adequate disclaimer of liability over the potential problems. This right of objection is not given over provisional sum expenditure, as the subject matter has been included in the contract with some definition of its nature. If the contractor found that the actual instructions were different from those inferred from the employer's requirements with detrimental effects on the design, he might well be justified in objecting, despite the lack of direct authority.

While the employer can instruct design changes, that is require the contractor to amend his design to suit the employer's revised requirements, the employer has no power directly to introduce design of his own into the project. This is one of the most efficient ways of confusing responsibilities, thus diminishing those of the contractor. If the employer has a facet of highly specialised work which he *must* design post-contractually, such as integral plant, he should ensure that the contractor accepts it with a categorical assurance that it does not affect the integrity of his own design. The employer cannot reasonably ask him to accept responsibility for the employer's design within itself. Even if the employer's design is for work to be performed by a separate direct contractor under clause 29 or by a named sub-contractor under the 'BPF supplement' (see Ch. 14), the same procedure is appropriate.

Except in this one unusual respect, which the contract does not consider, the contractor need not accept any design intrusion. The employer in turn should stay with his role of consenting to what the contractor now proposes or, if it is not too uninformative for the contractor, simply noting it. The exception is over changes of layout etc if some design amendment, forced upon the contractor in developing his scheme, is best dealt with by invoking such a change, which otherwise properly emanates from the employer. Here there must be agreement, and so approval.

Throughout the contract, any supplementary design needed to remedy damage to the works and for other reasons is part of the contractor's obligation to 'carry out and complete'. If the project comes to a premature halt due to termination of the contractor's employment, both clauses 27 and 28 provide for copies of design work up to that date to be handed over to the employer for use in completing the works. The contractor has no responsibility to produce anything more, whoever has determined. If work is completed normally, the contractor is responsible for providing as-built, maintenance and operating information under clause 5.5, but only to the extent specified in the employer's requirements or contractor's proposals. The supply of all working drawings under clause 5.3 will cover most points, unless there have been changes introduced on site without drawings. This is always a dubious procedure for reasons of sound communication, but is more excusable in this type of contract, although not necessarily more likely to occur.

Design liability: in general and the JCT contract

Following on from the discussion of the contractor's responsibility, there comes that of his liability when the design goes wrong. This is usually a contractual liability, and under the contract solely so, although the contractor may find himself with other liabilities and to other persons, as mentioned later.

Background liabilities

Consultant designers

A liability for design faults may fall into one of two categories: breach of a duty of general care or of a duty of strict care. The former is the duty of any designer, or other person exercising a skill special in character, to use reasonable care based upon the possession of the *normal* skill and competence expected of those in his position, as was set out in *Bolam* v. *Friern Hospital Management Committee* (1957). This was a medical and not a building case at all. It was quoted with approval in a civil engineering case *North West Water* v. *Binnie* (1988) over consultants' liabilities, when it was added: 'It is enough if [a consultant] exercises the ordinary skill of an ordinary competent man exercising his particular art.' Further and perhaps surprisingly, in the building case of *Hawkins* v. *Chrysler* (1986), it was held that while an architect was due to exercise reasonable skill and care, he did not have to warrant that materials used in a floor finish would be fit for their intended purpose. This would be subject to any express undertaking on his part.

The standard RIBA Conditions of Appointment in their very first clause say: 'The architect will exercise reasonable skill and care in conformity with the normal standards of the architect's profession.' An architect then, as an example, when entering into a normal engagement and for whomever he is working, is expected to be competent, prudent and so forth. Failure in these respects may constitute incompetence, undue error or negligence. But he is not necessarily expected to be the highest exemplar of his profession. This is sensible, as not everyone can be the best. By inference, some reasonable performances must be rather below average, although not too far. A client may expect a competently designed building, aesthetically and in layout (and must reject proposals which he can see do not indicate this), and also one which is technically sound in detailing and choice of materials, which he, as a client, may not be able to judge in advance. He may of course prejudice his position by requiring his architect to incorporate untried systems or materials and the architect, after duly registering his warnings, may not then be liable.

The duty of strict care goes further and extends to providing a design which will meet the criterion of fitness for purpose. Thus a designer who clearly undertakes to design a building which complies with express stipulations, perhaps by meeting given standards of performance, has a responsibility so to do. If his design is deficient then he is liable, even if it is a normally competent design and even if he has not been negligent. This is not a question of degree as to whether the design is better than normal. It is more an absolute question: has the specific standard been achieved? It is often the case that architects and others, by undertaking that some aspect of a building will have particular characteristics, accept a duty of strict care over that aspect, while retaining simply a duty of general care over the rest (see the *Greaves* case below).

Contractors in general

The general implied liability of a contractor, in the absence of special contractual conditions, goes rather further than general care. He has a responsibility to supply materials for the works which are reasonably fit for their purpose. This may be modified by the specification from the employer's side of a particular type or quality, which may increase or decrease the contractor's actual liability. The importation of 'purpose' gives a stricter degree of care than the general, while a precise specification (up or down) imports a narrower duty of strict care. Thus design and specification in detail by the employer's consultants, as is quite normal, places the contractor under a duty of strict care for all work so handled. Correspondingly they remove from him any duty or right to design or specify the elements concerned, which usually constitute the whole. Only if there is an element dealt with under such provisions as those for performance specified work in the JCT standard form (see Ch. 17), is some restricted area of contractor's design and specification introduced.

Design and build contractors
When a contractor erects a building which he has also designed, the matter of design must be added to that of materials and workmanship. In the case of housing provided in this way, there is an implied term at law of fitness for human habitation, strengthened by the Defective Premises Act 1972, and it appears correspondingly likely that terms may be implied in contracts for other building types, if such terms are applicable to all buildings of a type. There is therefore a prima facie situation that a contractor entering into a design and build contract not otherwise qualified has a greater design liability than a consultant who performs design alone, at least when engaged by the employer. While the actual situation under the JCT form with contractor's design is yet to be considered, law and cases illuminating the foregoing principles may be taken first and indicate some continuing uncertainty over the precise law in this area.

The Sale of Goods Act 1893 enacted that those selling in the course of business to consumers imply terms over *goods* about conformity with description or sample and quality or fitness for a particular purpose, if reasonably knowable by the seller. These provisions have been re-enacted in the Sale of Goods Act 1979, while the Unfair Contract Terms Act 1977 has rendered unenforceable many exclusion clauses in contracts. Between the first statute and the latter two, in the case of *Young & Marten Ltd* v. *McManus Childs Ltd* (1969) the House of Lords held that similar terms to those over goods were to be implied in contracts for work and materials. The key feature of an implied fitness for purpose of work, as given in *Corben* v. *Hayes* (1964), is that the employer should be relying upon the contractor's skill and judgement and not upon his own or those of his agents.

Two cases distinguish this position. In *Lynch* v. *Thorne* (1956), the builder contracted to complete a partly built house in accordance with drawings and specification (apparently his own), which featured a one-brick external wall. This wall was not weatherproof, even though soundly constructed, but the Court of Appeal held that the express specification overruled any implied warranty of fitness for habitation at this point, so that the purchaser's action failed. The specification, it may be noted, was not an exclusion clause. By contrast, in *Hancock and Others* v. *B. W. Brazier (Anerley) Ltd* (1966) the builder was held liable under a contract entered into rather similarly. Here hardcore was shown under the floors, but the precise material was not specified, so that the implied warranty remained in this respect. Its chemical content, unknown to the purchaser and also to the builder, caused failure in the floors. Lord Denning in the Court of Appeal commented in essence that a builder who contracts to build a house implies three terms: execution in a good and workmanlike manner; use of good quality materials; and reasonable fitness for human habitation. He was to repeat these comments in the case next mentioned.

This is the case of *Greaves & Co. (Contractors) Ltd* v. *Baynham, Meikle &*

Partners (1975) relating to a design and build contract. Here the contractor sought to recover from his structural consultants the cost of remedial work, for which he admitted liability and so he had to perform it free for the employer. The first floor of the steel-framed warehouse was to be formed in precast concrete planks and *in situ* filling. It was known to the consultants that forklift trucks were to run over this floor, carrying heavy drums of oil. British Standard Code of Practice 117 contained a warning note about the effect of vibrating and moving imposed loads on such a floor. The technical details in the case are interesting but are not relevant here. It was found that the defects were due to deficiencies in the consultants' design, rather than in the contractor's workmanship, as had been contended.

The judge applied the *Bolam* principle and held that there had not been a failure to exercise reasonable care, so that the consultants were not negligent. But he held that they had failed to meet an implied contractual warranty that the design would give a building fit for its purpose of accommodating the working trucks, and so would still have been liable in the absence of negligence. On appeal, Lord Denning stated there had been a special warranty of fitness implied *in fact* in this instance, so that it was not necessary to answer the secondary question of whether a more general duty was also implied *in law* for the consultants, that is, to meet 'fitness for purpose' rather than just exercise 'reasonable care and skill'.

Several comments may be added. No question was adduced before or during the case that the contractor would not be liable to the employer if the defects proved to be in design, rather than in workmanship. Either way in design and build, the contractor had a liability. Second, and at least in cases of special cases of knowledge of use of the building, a consultant has a duty to provide a fit design and is liable if he does not, without the necessity of demonstrating his negligence, as is the case over his undoubted duty to use reasonable care and skill. Lastly, when the consultant is designing for the contractor in such cases, he is liable to indemnify the contractor over the cost of satisfying the employer's right to have faults due to poor design rectified.

The case did not have to deal directly with the liability of contractor to employer over design in a design and build contract, and it stopped short of a rounded treatment of the extent of a consultant's liability. The next case also stopped short in its approach from a somewhat different direction, although all these cases suggest definition of a deep degree of liability.

The case of *Independent Broadcasting Authority* v. *EMI Electronics and BICC Construction Ltd* (1980) was also a design and build situation, in which EMI contracted to erect a television mast. BICC were their nominated sub-contractor and designed the mast which was of advanced design, the full implications of which had not been researched in the industry at the time. BICC assured the Authority *after* obtaining the sub-contract, and while designing, that the design was satisfactory. The mast collapsed during a gale, while in use. It was held that the collapse was due to vortex shredding and asymmetrical ice-loading negligently not taken into account in the design, which was a contractual responsibility.

The House of Lords agreed with these points. It was also held that there was an implied warranty by EMI to IBA of fitness for purpose. EMI were in breach of this warranty, and BICC (the actual designers) in turn were in breach of their sub-contract with EMI. But further, BICC were liable to IBA *in tort* because they had negligently assured IBA that the design was adequate. They had a duty of care towards IBA, who had relied on their expertise and assurance, on the principle in *Hedley Byrne* v. *Heller* (1964) of duty to use care when another would rely on advice given. This duty existed apart from any contractual relationship via the main contractor, which relationship did not cover the assurance given post-contractually.

As there had been negligence in the design of the mast, so offending against the criterion of reasonable care, it was not essential to prove a breach of the fitness for purpose criterion. However, one of their Lordships did say in passing (that is as *obiter dicta*) that the greater obligation of meeting fitness for purpose applied to the supply of an article, unless expressly or impliedly negatived by the contract, rather than the lesser obligation of reasonable care normally applying when the contract is for a design service alone. Another of their Lordships expressed the view that he would be surprised if indeed it were to be found, if the necessity arose, that the design was not inadequate for its purpose. The House therefore did not have to resolve finally the fitness for purpose question, although it looks fairly clear how they would have done it had they needed to do so.

An Irish case, *Norta Wallpapers* v. *Sisk* (1978), looked at another sub-contractor situation. Here again the sub-contractor was nominated and designed and installed a roof, but the situation was distinct. The contractor had had to accept the nomination and the design introduced by it. He therefore had no liability over design faults and there was no warranty of fitness for purpose to be implied in the main contract. This case may be noted in respect of nominated sub-contracts as discussed in Chapter 18.

In the case of *Basildon District Council* v. *J. E. Lesser (Properties) Ltd* (1985) a term of design for fitness for habitation or alternatively of strict skill and care was implied in the contract by the court. This was not a design and build contract in form but a JCT 1963 form, and the recitals stated that the drawings had been prepared under the direction of the council's engineer or architect. The contractor, as a regular system builder, had in fact first supplied the drawings which then became the contract drawings. He was held to have 'a continuing responsibility for the design' which had been embodied as it stood into the contract. There was reliance on the contractor 'to produce habitable dwellings'. Although this was not a design and build contract, the known purpose led to the implying of the strongest level of liability over design proferred. There was a direct statement of strict liability, without expansion of the principles, in the case of *Viking Grain Storage* v. *T. H. White* (1985), over a design and build contract which did not track back to the consultant, but where the dispute was solely between the employer and the contractor.

Strict liability over statutory requirements of the contractor to the employer in the specific question of breach of the Building Regulations was held to lie in

Newham LBC v. *Taylor Woodrow* (1981), the well-known case of the progressive collapse of a series of flats after a gas explosion in the Ronan Point block.

In summary, these cases appear to support a number of salient points:

(a) A contractor who designs and builds has a contractual obligation to meet fitness for a known purpose, unless he excludes it from the contract. This is a greater obligation than that normally assumed by a person who designs for an employer, who in turn engages a contractor solely to build.
(b) A sub-contractor who designs and builds work has a similar obligation to the contractor, who is liable to his employer.
(c) A consultant engaged as a sub-contractor to design work forming part of a design and build contract also has this higher obligation to a contractor, provided he is aware of the contractor's obligation to his employer.
(d) If an employer has cause to rely on the exercise of reasonable care by a sub-contractor of any type in a matter of design, the employer may have a right of action in tort over negligence in that exercise. Unless specially qualified, this duty does not extend to fitness for purpose, where negligence need not be shown.

The question of tortious liability is taken further in Chapter 10 under 'Whether liability is in contract or tort'.

There is a strong tendency for employers to seek to impose fitness for purpose by such means as noted below as the modification or outright deletion of JCT clause 2.5.1 when this might apply. It may also come about by the imposition of an onerous collateral warranty (see Ch. 10) in favour of some person or persons. Contractors should be alert to these happenings and to whether they are reasonable in given circumstances. Even from employers' point of view, they should be viewed against the possibility that at least the smaller contractor could be put out of business by a large claim, so that the employer has no line of redress open to him, even against sub-contractors. Insurance, discussed hereafter, is not usually available to the contractor against fitness for purpose claims and so will not protect the employer either.

Extent of liability under the JCT contract

The statement over the contractor's design warranty and liability in the JCT form is given in clause 2.5 which is discussed as a whole later in this chapter. The present section relates parts of clause 2.5.1 to discussion in preceding sections. Four expressions call for comment:

(a) 'Insofar as the design . . . is . . . in the Contractor's Proposals': if design is in the employer's requirements, it is not part of the contractor's responsibility unless he is made responsible for checking it before basing his own proposals upon it. The desirability of the contractor taking all responsibility and the possible problems in asking him to do so over some aspects have already been discussed. The clause is thus allowing for the problems, but is also a warning here that the employer should clarify the allocation of responsibility, otherwise he may retain more than he intends.

(b) 'any defect or insufficiency in such design': this term equates with the results of not exercising reasonable care, when read with that under (c).

(c) 'the like liability' to that of 'an architect or . . . other appropriate professional designer . . . competent [and] acting independently under a separate contract': the designer's liability is that of reasonable care, so that he is liable for negligence but not for failing to design a building that is fit for its purpose, unless this is known to him quite explicitly. Such a designer is described as 'competent', with the overtones of being skilful but not necessarily outstanding. He is acting as 'supplier of the design' to which another is to erect the works.

(d) 'liability . . . whether under statute or otherwise': from earlier discussion it may be seen that most areas of liability are 'otherwise', being defined by decided cases. The duration of liability is discussed in Chapter 10.

Clause 2.5.1 thus reduces the liability of the contractor from that which usually rests upon a builder who designs and builds to that of an independent designer. In this respect it may be viewed as an exclusion clause, although whether it would fall foul of the Act is doubtful. The contractor is essentially liable for negligence and failure to perform his design and, as always, for not providing workmanship and materials of contract quality.

Design may be included in the employer's requirements but fall outside the 'Insofar as . . . comprised in' term of the clause, so reducing the contractor's potential liability. It is, however, possible that the employer's requirements may include stipulations that aspects of the contractor's design shall meet 'fitness for purpose' criteria, so seeking to increase potential liability. Such stipulations will be invalid if they 'modify the application or interpretation of . . . the Conditions', as clause 2.2 has it. To avoid this, they should not attempt to say anything about clause 2.5.1, such as 'the normal design and build liability is substituted', but simply spell out what in particular the contractor must achieve in his design, perhaps by a performance specification. If they do this they will still be placing on the contractor a 'like liability' to that of an independent designer who may, as discussed, assume an additional responsibility in just such a way. It would appear that the stipulations will then be valid and effective. The contractor, for his part, should look carefully at such provisions when assessing his responsibility.

Some employers take a more simplistic approach to clause 2.5.1 by seeking to delete it completely, so reinstating fitness for purpose for the entire works. Any contractor should be alive to what is being done and react accordingly. Unless the purpose is very clearly defined, the deletion may fail at least partly in its aim. It is unlikely that the contractor's consultants will accept a stepping-down of the action, while insurers do not normally offer appropriate cover.

Whatever responsibility the contractor may assume is not reduced by the wording of the third recital in the articles of agreement: 'the Employer . . . is satisfied that they appear to meet the Employer's Requirements'. The employer should have dealt with all divergences in the contractor's proposals, either by seeking an amendment to the proposals or, as the footnote advises, by bringing his requirements into line with any acceptable divergence.

Clauses 2.5.2 to 2.5.4 come under a side heading of 'Limit of Contractor's design warranty': they define the limit, rather than limit the liability. Their detail is discussed later in this chapter. Clause 2.5.4 states that design which the contractor sub-contracts is included in all reference to design in the conditions. This appears *inter alia* to make the contractor liable over design faults of a sub-contractor, as would be expected from the cases already considered. If the optional Sub-Contract DOM/2 is used, which the employer cannot insist upon, it makes the sub-contractor liable to the contractor in similar terms to clause 2.5.1, although not to the employer.

However, the expression 'like liability' in clause 2.5.1 needs comment again in this respect. Under the RIBA Conditions of Appointment, clause 3.8, the architect is not 'responsible for the competence [etc]' of a sub-contractor or other person, who himself bears the responsibility for any design he produces, although the architect is responsible for checking that the person has produced a design suitable in principle for the scheme. This applies when the employer either has nominated the person to design or has accepted the architect's nomination. JCT clause 18.2.2 allows the contractor to sub-let design only with the employer's consent, so that an equivalent position may be held to result. It may therefore be argued that the contractor's 'like liability' excludes liability for a sub-contractor's default, at least for architectural design. Clause 18.2.2 states that its implementation is not to 'affect in any way' the contractor's obligation under clause 2.5. Leaving aside any distinction between 'obligation' (not mentioned in clause 2.5) and 'liability', this statement begs the question of what clause 2.5 provides and so of what is unaffected.

In the absence of judicial decision on this point, and cases have been decided unexpectedly on narrower issues, the employer is advised when consenting to a sub-letting of design to stipulate that it is on condition that the contractor remains responsible for the design as though it is his own. There is no employer/sub-contractor agreement available as with other JCT forms (although see the mention of tortious liability under the next heading), and safety is preferable to sorrow in an uncertain situation. Such a post-contract stipulation does not fail because of clause 2.2, although it might be held to constitute an unreasonable withholding of unqualified consent.

A distinction may be intended in the reference to design which the contractor causes 'to be prepared or issued by others'. 'Prepared' relates to design resulting from sub-letting. 'Issued' carries the meaning of design already performed and taken, as it were, off the shelf for use in this particular contract. This may apply to a standard structural system which a sub-contractor puts forward in response to a specific need, and here there will be a proper sub-letting and the resulting liability cannot reasonably be distinguished from what has been discussed. At the other extreme lies the design incorporated in any artefact: this is outside the present scope and is governed by the usual considerations applying to goods. Between there are cases of design 'issued' apart from the work itself, say in data sheets, and on the basis of which the contractor himself chooses precisely what to do. In these cases, the contractor is designing by selection and so is liable to the employer, whatever redress he may retain against the issuer of the design.

Subsidiary points over liability

There are several important subsidiary points in relation to the main question of the contractor's liability whether under the JCT form or not. They are logical consequences of the preceding discussion, where some of them have already been mentioned.

Where the contractor engages an outside design consultant, he should check that the consultant's potential liability to him matches his own to the employer. Under the JCT form as it stands, the contractor has a 'like liability' to that of an independent consultant, so that the regular terms of engagement are suitable. Under a design and build contract embracing 'fitness for purpose', including a JCT contract which introduces it by performance specification or otherwise, the contractor must seek a deeper commitment. This may be to meet some particular standard in an element of a building, as was held to have happened in the *Greaves* case. Especially in a non-JCT contract, a blanket acceptance of the full obligation over design in an unqualified design and build situation may be needed. The consultant in his turn needs to be watchful as to the responsibility that he is assuming.

The limitation period during which actions may be brought over residual liabilities is discussed in Chapter 10. The extent and duration of these liabilities are affected by the nature of the contract, while there are also differences according to whether the action is in contract or tort.

Consultant and contractor alike should check their insurance protection in these circumstances, in view of the possibility of very heavy consequential loss to the employer. For the consultant, the wider scope of liability is the major issue. For the contractor, it is the possible duration of liability that goes with the responsibility for design. JCT Practice Note CD/1A considers this matter. It indicates that the normal professional indemnity policy covers matters within the 'like liability' category, broadly negligence, and does not extend to cover 'fitness for purpose'. Equally it does not cover materials and workmanship. The extent of cover which it is possible for either person to obtain in the insurance market is likely to fall short of the maximum need that can be anticipated. The note mentions also the position of a contractor who is involved in design and build work only occasionally and so does not maintain an annual insurance policy against the risks involved. Any contractor engaging a consultant should check the indemnity secured by that person's policy against the risks concerned.

The JCT form contains no requirement for the contractor to insure, even over his negligence in design, that is the 'like liability' category. This is in contrast to the GC/Works/1 form discussed in Chapter 15. The employer should reasonably make a stipulation in the employer's requirements about what insurance he expects, both in its nature and the amounts covered. He should also introduce rights to check that the insurance is being maintained and, if not, to be able to insure and recover costs, on the lines of what is in JCT clauses 21 and 22. It is no consolation to the employer if the contractor does not have the cover and is driven into insolvency without being able to meet his obligations to the employer. Insurance is expanded upon under the next heading.

Design insurance

All contractors must carry public liability insurance as a statutory requirement and it is usually also a contract requirement. This insurance covers liability arising during or consequent upon acts of construction including such matters as negligence during performance. It does not cover liabilities to replace defective materials and workmanship. Of particular importance for present considerations, it does not cover liability for design matters, a provision not needed by contractors who do not engage in design and build activities.

Contractors properly need insurance protection in respect of design responsibility which they assume, even at the level of quite minor works or when developing the architect's design or giving the occasional comment to help out, although it may be suspected that many have none. However, for present purposes it is primarily the full design and build situation which is in view, whether undertaken for the complete project or for a distinct portion, as for instance with the designed portion supplement. The case of design development may be just as vital, when it assumes the scale which is present in such an approach as that of the BPF system (see Ch. 13). It will need care to ensure that the insurance covers what is required, as there is the question of potential overlap between design provided by the employer and that produced by the contractor. This should normally be accounted for not by special insurance terms, but by clear definition in the technical documents of who is responsible for what and particularly of whether the contractor is to check and take responsibility for the employer's initial design. In any doubtful case the insurer should be consulted.

The sector of insurance concerned is that of professional indemnity insurance as routinely used by consultants, and the very name indicates why a contractor may overlook its applicability to his own business of building. The contractor should take into account that the insurances of his consultants and others will provide only a standard level of cover, as discussed below, and that he may be unprotected in vital areas if he accepts liabilities not covered by their insurances and his own.

The protection afforded by a policy is affected by the nature of the business being insured. Companies may go out of existence by such routes as liquidation and the redress available may go with them. Partnerships carry on beyond the presence of a retiring partner (who may still have liabilities to and through the partnership) and until themselves dissolved, when the liabilities of the various partners in relation to clients may survive. Individuals, including ex-partners and sole traders, are subject to valid claims until bankruptcy or death. In general though, professional indemnity insurance is hardest to come by for a company.

In what follows, discussion is essentially limited to contractors, as professional consultant designers are regularly involved in the insurance market concerned.

Obtaining insurance

The JCT with contractor's design form makes no reference to professional liability insurance in its design responsibility or insurance clauses. This is done only in the BPF/ACA and GC/Works/1 contracts out of those discussed in this book. It is a type of insurance which is expensive to obtain, as the level of claims has risen exponentially with the introduction of modern and untried forms of construction, the increased numbers of more aware clients and the erosion of some of the aura surrounding professional consultants. It is nevertheless highly important to the contractor to be properly covered as the value of claims can be very high, enough to drive him into insolvency without insurance. For the same reason, the contractor's insurance is important protection for the employer, who will otherwise be left uncovered to a degree, perhaps considerable, if the contractor does become insolvent.

Understandably there may be detailed stipulations about professional indemnity insurance in the employer's requirements, but the problem may be that more is asked for than can be obtained in the insurance market. In practice it is difficult to obtain cover for every area of risk and even for a policy to be written which will offer such cover. In particular, the increased use of collateral warranties has introduced extra avenues of liability (see under 'Collateral warranties' in Ch. 10), it may be suspected beyond what the drafters envisaged in some cases. However, as wide and deep cover as is available should be sought, subject to premiums not being absurdly high.

There are two basic approaches open to the contractor, who faces potential liabilities during construction and most particularly afterwards for a large number of years (see under 'When liability is extinguished' in Ch. 10). If he performs only the occasional design and build scheme, there is the option of insuring each project by a single premium on a once and for all basis until the expiry of liability, but this is unpopular with insurers and so quite difficult to obtain. If the contractor is regularly engaging in design and build work, it is more satisfactory for him (and probably all that he can obtain) for there to be an annually renewable insurance to cover claims brought in the current year, whenever they were caused. Even if he ceases to engage in design and build work, it will then be necessary for him to continue insurance on a 'run-off' basis, that is for such a period of years as remains until his potential liabilities are extinguished. How an employer can ensure that this run-off insurance is maintained is a near-intractable problem.

It may be possible for a contractor when first entering into professional indemnity insurance to obtain retroactive cover, that is for work performed before the commencing year of insurance. There may be special provisos attached to such insurance and certainly enquiry about whether any claims are pending or any causes of claims are suspected. A contract of insurance is one of utmost good faith and the proposer is under an obligation to declare all material facts. However, it will be very useful, for instance, to obtain such cover for design work undertaken on a scheme ahead of any certainty that a contract is to be awarded.

There are two approaches to the way in which limits of liability are set on an annual basis. One is by covering an amount as for 'any one claim and in all', which means that an aggregate amount is covered, so that any one claimant may receive a reduced amount because of other parallel claims. The other is for an amount for 'each and every claim', so that every claim receives its full amount in settlement up to whatever is the limit of indemnity. Naturally, aware employers prefer the second approach. It is possible to arrange policies to give both forms of cover, according to client type. As usual with insurance, policy excesses are available or compulsory.

Policy terms

There are a number of considerations additional to how often the contractor engages in design and build work, how much work he performs each year and how he is likely to buy his cover.

As well as the statutory periods of liability, it is necessary to take account of any specially imposed periods in a particular contract. This is likely to be a particular feature of any collateral warranties, which are a substantial area of interest to insurers in view of these and other terms imposed beyond the design and build contract itself, as well as the range of potential claimants. In many cases, insurers are not prepared to underwrite risks which spawn a well-nigh indefinite number and variety of claimants and run on for very long periods. It is possible to conceive situations in which the statutory periods may be far outrun. A contractor needs caution and advice on all aspects (of which insurance is but one) before entering into such onerous warranties, if he does so at all. His only pale consolation then may be that if he is not properly insured and fails, the person with the warranty in his favour is unlikely to obtain recompense!

The regular wording of policies is such that they provide cover on a normal skill and care basis only, that is against ordinary errors, omissions and negligence, and not on a strict care or fitness for purpose basis. This meets the standard JCT case in clause 2.5.1 (which may have a limit of value introduced by the appendix, as referred to by clause 2.5.3), but does not cover the higher standard of clause 2.5.2 over dwellings. It will need an extension to the policy to cover these higher levels, and usually (again) insurers are extremely unlikely to accede to this. They may well specifically exclude by policy terms such liabilities, even if only likely to arise occasionally and then sought to be covered by a supplementary premium.

The contractor will usually incur several responsibilities which are to be observed closely in the event of a claim. As elsewhere in commercial dealings, he must act to avoid or mitigate any loss or potential loss and not remain simply passive. His extra costs of doing so will then rank as part of his claim.

He must also give early notice of a possible claim: this is an onerous stipulation and requires a degree of openness on the part of the contractor during progress when he may be hoping that he can take avoiding action so that no claim arises, such as by compensating redesign of a later part of the building. In such an instance of potential claim and when there is a clearly established claim, the contractor faces a further question. This is that once a claim arises, the insurer has the right to take over the negotiations for its settlement. In the context of design and build, this effectively means that the insurer enters into any redesign discussions and there may be some resultant compromise of the contractor's design if work is still proceeding.

It is possible that the insurance contract may be voidable by the insurer if unapproved alterations to the works are carried out or if there is a lack of adequate maintenance or repair. Neither of these contingencies is within the control of the contractor, but equally they are not matters which are likely to concern him unduly, provided he can demonstrate adequately that he was not involved. However, they are issues of considerable concern to the employer or to others who may acquire an interest by means of a collateral warranty.

Other problems may arise over non-disclosure of the existence of warranties at all and over mismatches with the standard inclusions and exclusions in the professional indemnity policy. It may be possible and acceptable to have endorsements on the policy to cover at least straightforward warranties. Care will be needed to ensure that all of these issues are adequately dealt with and carried forward into future years if there is a change of insurers.

Consultants and sub-contractors

The position is more complex when design is let out to consultants or sub-contractors, who need protection by their own professional indemnity insurance. The legal status of different businesses has been mentioned at the beginning of this insurance section.

Consultants are bound by their own professional bodies' regulations to carry insurance, which will be on a skill and care basis corresponding to their standard terms of engagement, but the existence and level of cover should be checked. Care will be needed in checking whether consultants who are not corporate members of an appropriate body are insured. It will be similarly necessary to check on the position of sub-contractors performing design, especially if they do this only irregularly.

The contractor's policy often does not require insurance by consultants and sub-contractors, but there may be a lower head premium if they do. In any case there must not be any erosion of the insurer's right of subrogation against such persons, that is to seek recovery of amounts paid out under the policy from them. Prudently, they should insure against their own liabilities arising in this way. It is possible to obtain an extension of policy to cover the contingent liability relating to a consultant or sub-contractor who goes out of business or otherwise cannot meet a claim, or to the situation where allocation of responsibility between persons is difficult. This may be valuable, despite the fact that or perhaps because the policy does not insist on such persons insuring for relevant work.

Design copyright

No legal difference arises because of the design and build principle itself. Copyright may be distinguished from a patent right: the former is a general right over artistic quality conferred on its owner automatically by statute, while the latter is a particular right over function and manufacture, obtained by application and defined procedures, although again under statute.

General background

The present statute is the Copyright, Designs and Patents Act 1988. This applies copyright to *work*, rather than *ideas* not embodied in some material form of work. The work must be original in the sense of originating from its author, rather than necessarily being novel or innovative, and be the product of skill and labour. The Act defines 'artistic work' so as to include drawings, plans, models, buildings and structures, among other things. These need not have artistic quality and need not be finished work. Even this compressed outline indicates that work produced for normal architectural and related purposes is covered. Written work also falls within the scope of copyright.

Copying may be of a building by means of drawings or by another building, or of drawings by building or by other drawings. All four of these constitute an infringement of copyright, except copying a building by drawings as in a measured survey. Photography of a building is not an infringement, although the use to which it is put may be. Photocopying of work may be an infringement. Legal cases in this area have produced some hard-argued and finely decided points beyond present concerns, including consideration of how close the copy must be, to be an infringement.

Finished design

When a design and build contract has been entered into, the employer obtains ownership of such drawings as he pays for, just as he comes to own the building, and so its design. Except by special agreement, copyright in the design as contained in the drawings and building remains with the designer. By virtue of the contract, the contractor is obliged to produce the building in accordance with the drawings. He may fail to produce it, or be rendered unable to produce it, in the circumstances of determination of his employment by the employer or by himself. If so, he still retains the copyright. However, the employer may reasonably proceed to complete the building on the basis of the drawings, if he is so able. The

JCT clauses are not explicit here; in both clauses it is provided that drawings etc prepared up to the date of determination become the property of the employer. This serves no purpose if he cannot use them. If it is the employer who determines, he has the further right expressly stated of *completing* the design and construction, so that he may use the drawings already prepared but not used.

If the contractor retains the copyright then, as indicated, he has protection against unauthorised use of the design contained in his drawings and building. Even so, ownership of the copyright does not carry with it the right for the contractor to reproduce the design in its main substance in another building, except with the present employer's agreement. The reutilisation of details is quite permissible. If the contractor is providing some standard type of building of his, so that reproduction in the present contract or in future work is inevitably substantial, he is advised to make sure that the employer accepts that his building will be by no means unique.

In the case of mass-produced components, it is usual for the designer to register the 'design copyright', a matter of industrial design rather than artistic character. This is wider than normal copyright in its effects and is somewhat akin to obtaining a patent, and is in fact registered at the Patent Office, as it restricts the right of reproduction etc to the holder, even against someone who independently evolves the design. If the contractor develops components as part of his present design and envisages using them in future projects, he may be advised to seek registration so that he is protected from copying, over and above his own right to produce.

The JCT clauses most obviously having some bearing on copyright and discussed elsewhere are:

Clause 5.6: Restrictions on the use of documents by the party receiving them.
Clause 9: Indemnity to employer over infringement of copyright etc held by third parties.
Clause 27.6.1: Provision of drawings by contractor, retained by employer after determination.
Clause 28.4.2: Provision of drawings by contractor, retained by employer after determination.
Clause 27.6.2: Payment of others by employer to complete, incorporating design to date.
Clause 28.4.3: Payment for drawings by employer, with possibility of using them to complete.

Unfinished design

A twilight position exists during design development, especially when more than one design is under consideration, as pursued in Chapter 6. Clearly the employer cannot take even one sketch drawing produced by a tenderer and pass it to another saying 'use this in your scheme', for even a sketch or partly prepared drawing has copyright protection as there is expressed work based on skill. On the other hand, ideas not given material expression are not protected. The employer may therefore be within the letter of the law to ventilate one tenderer's ideas with another as much as he wishes, so securing cross-fertilisation of ideas. Within limits a drawing conveys ideas. So, while he could not pass on a fenestration sketch, he could well suggest to the other tenderer that he run the building north and south like the first tenderer, rather than east and west. This could be of major commercial significance, but is not what copyright can protect. Between these examples, lies the question of internal planning, where copying might range from precise reproduction with room sizes and door positions, to some repositioning of the boiler room to give a better clustering of service ducts. One might be actionable, but not the other. Legal immunity may here run counter to ethical probity.

The employer is looking for the best design, and design and build has the advantage of offering several designs, with the possibility of comparison during development. He can hardly insulate one design from the other in his mind, so that he is bound to carry ideas over from one to the other. However, it is no defence against infringement of copyright to plead that a part of one design was abstracted as an *idea*, which was then injected into another design, so that *copying* did not occur.

The contractor when tendering in progressive competition is in a difficult position: he must declare his scheme to gain consideration, but before he is assured of success. He may thus risk 'borrowing' by others, consciously or otherwise, probably with no remedy. It is therefore highly desirable that the employer should give a categorical assurance of confidentiality when proceeding by stages, clearly defined or otherwise. This assurance should apply to a tenderer's scheme while it is being considered, whether it is accepted or not.

Documents and obligations: the JCT contract

The clauses, etc. taken here all bear on the foregoing discussion, although they also have significance at later stages of practical activity. Clause numbers in brackets are those of the anticipated revised JCT section headed contract (see also the comparison table in the Index of JCT clauses).

JCT employer's requirements, contractor's proposals and statutory requirements

These three elements constitute the definition of what the parties have agreed shall be produced by the contractor ('the Works') in return for the contract sum, which is discussed in Chapter 6. The first two are introduced throughout the articles and clauses as what has been explicitly agreed, while the last underlies the agreement without being spelt out.

The 'Articles of Agreement' are strictly a distinct contract document from the

'Conditions', which contain the clauses of the contract. Major points in the articles, and in relevant clauses are taken here.

JCT recitals to articles of agreement

The recitals contain spaces for insertion of a brief title on the works and their location, followed by standard statements of events up to the formalising of the contract. Important points here are:

(a) The employer has issued his requirements for the works which he requires. There is no elaboration over precisely how he has defined his requirements.
(b) The contractor has responded by submitting his proposals for carrying out the works defined by the employer. It must be inferred until article 1 is reached that design is included. There is reference to 'the sum' payable here, taken up in Chapter 6.
(c) The employer has examined the proposals and 'is *satisfied* that they *appear* to meet' his requirements (emphasis added). The emphasis highlights a rather odd statement, at first sight suggesting that reality is far away, although the employer might not think so, while the contractor might rather not have the 'appears' there at all. The purpose of the wording is to cover the as yet uncompleted obligation of the contractor to produce complete designs and specifications to meet the employer's requirements adequately over function and quality. It is being stated, so the JCT practice note avers, that the requirements and proposals are not in conflict so far as they go, but without this relieving the contractor of his responsibility to perform his contract. It would be less likely to lead to legal controversy if something like the explanation in the practice note formed part of the recital.
(d) The inference under (c) is that there is no divergence between the requirements and the proposals. If in fact there is, the employer 'appears' to have signed away any right to a correction being made automatically and without charge by his expression of satisfaction, so that the proposals prevail as the contract statement. The footnote [b], which is not part of the contract, recommends that any divergence knowingly accepted by the employer as part of the contractor's proposals should be dealt with by bringing the employer's requirements into line. The possibility of discrepancies within either set of documents is dealt with in clause 2.

JCT articles of agreement

Relevant points for the present discussion are in articles 1 and 4.

Under article 1, the contractor agrees to complete the design, so taking affairs beyond the stage of the employer's requirements and the contractor's proposals, recorded in the recitals. This is additional to the other element of carrying out and completing the works, which is the sole element in other JCT contracts. Both elements are repeated and elaborated in clause 2.1.

Under article 4, the requirements and the proposals (along with the contract sum analysis) are to be signed and identified in appendix 3. This requires a list of drawings, specifications, schedules and so on. This statement and the mention of the same documents in clauses 2.1 and 2.2 in conjunction with the articles, conditions and appendices, are the nearest approach there is in the contract to an explicit list of contract documents.

JCT clauses 2.1 to 2.4 (1.2, 1.5 and 1.7) – Contractor's general obligations

Responsibilities

Clause 2.1 reiterates the contractor's responsibility under article 1 to 'carry out and complete the Works', which he achieves at practical completion under clause 16.1, although his residual liabilities extend well beyond there. On the one hand this obliges him to perform the whole works and not break off after just some part, but on the other hand allows him discretion over *how* he performs it in terms of sequence and general planning of construction operations. Here, and wherever there is more than one way of producing a given end-product, such as concrete of a defined strength, the contractor may exercise choice unless the documents restrict him.

This is all quite normal in building contracts. Where the present clause goes further is over completion of the design by the contractor and 'the selection of any specifications . . . so far as not described or stated in the Employer's Requirements or Contractor's Proposals'. This gives the contractor the duty and the exclusive right to make the selection. It is therefore incumbent on the employer to be specific over any materials or workmanship that are critical to his required building. Further, if he leaves the choice of which material to use (as usually he will) to the contractor in any situation, he should ensure that any limit is sufficiently defined. This need not be by absolute identification: it may be adequate to specify some performance factors such as load, deflection, thermal transmittance or surface characteristics.

Missing from clause 2.1 are words equivalent to those about qualities and standards being reasonable when left to the opinion of the architect, which occur in other JCT contracts. The employer's requirements, mentioned here (but not the contractor's proposals), may still refer to such cases and the opinion of the employer. If so, the same position over residual liabilities will apply in that the contractor will be relieved in these cases. The point is considered further under clause 30.8.1. The lesson for the employer is that he should not indulge in this form of words without strong reasons.

Divergences

Clauses 2.2 to 2.4 deal with clashes in the documents, rather than between them and the parties' expectations of them. The first provision, in clause 2.2, is for the three parts of the standard contract form to take precedence over the specially prepared requirements, proposals and analysis. This is the regular JCT approach, which reverses the usual legal rule of interpretation that the specially prepared overrides the standard, as expressing more closely the intentions of the parties. The JCT principle is obviously useful in nullifying accidental divergences introduced into the contract documents. But it also acts against the deliberate inclusion of clauses, probably in the requirements or proposals, which are meant to 'override or modify' some standard contract provision. This is particularly likely to crop up in the design and build field, where procedures tend to be more flexible than under more traditional arrangements. It needs to be done with care and advice, as such actions can produce unexpected effects, due to the interaction of clauses. In general, it is best left to cases in which some amplification is required, although even this should be done cautiously, because again the unexpected can occur only too easily.

If it is desired to draft in a change, this should be done by direct amendment of the text, as well as any statement of intent given in the other documents. The deletion of clause 2.2 itself would give unlimited licence, but may also lead to unlimited side-effects with the specially drafted documents then being uncontrollable.

Clause 2.3 deals with any divergence between the employer's requirements embodied in the contract and the definition of the site to be given post-contractually under clause 7. Such a definition could have been provided in the employer's requirements or, again in advance, by adequate marking on site. Any error here may affect the contractor's access or working areas or, more drastically, the positioning or even possibility of the works as designed. In any case, the procedure is to give a resolving instruction and so introduce a change (the term replacing 'variation' in this contract) under clause 12, with a consequent adjustment of the contract sum.

Discrepancies

Clause 2.4 takes up discrepancies within the set of documents produced by one part or the other. It operates on the *contra proferentem* rule that, in the event of doubt over interpretation, a document in a contract is to be construed against the party offering it. Thus in the case of a discrepancy within the employer's requirements under clause 2.4.1 it is the contractor's proposals which prevail, on whichever of the discrepant requirements the contractor based his proposals. The amount payable is also unchanged. The implicit assumption is that the contractor did not notice that there was at least one more option available, and equally that the employer was satisfied with the one selected, since the proposals 'appear to meet' his requirements, as stated in the third recital of the articles. It is to be hoped whenever a discrepancy of this type is noticed pre-contractually that the party doing so will clear it with the other at that stage.

More awkwardly, clause 2.4.1 extends the employer's requirements to include changes under clause 12, with the contractor's proposals again prevailing. Clause 1.3 defines both the requirements and the proposals in terms of the recitals, that is as documents finalised when the contract is formalised. Strictly, therefore, the requirements cannot 'include' post-contract changes. Even if this is overlooked, how can the contractor's proposals prevail over what is in a change? This may be intelligible if some part of the change instruction does not match up with the contract content, that is it seeks to change what is not there, although this is not properly an internal discrepancy as envisaged. Otherwise it means that the change becomes of no effect or, even worse, that the contractor may act on part of it to the employer's detriment, whereas what is needed is a provision for resolving the impasse. Perhaps the employer has to go away and rewrite his change instruction, so that the contractor's proposals do not automatically prevail. Hopefully, the need to give notice will lead to a rational outcome here.

In the alternative situation of a discrepancy in the contractor's proposals under clause 2.4.2, the end-result is still that the non-proffering party (in this case the employer) has the advantage. The difference is that the contractor has the opportunity to propose an amendment to solve the problem in the first place, which is reasonable enough when it involves *his* design. Indeed, without an amendment the design may be unworkable. The employer may well consider it fair to accept this, just as he accepts a development of the design. However, he need not do so even here, and indeed may decline, which is all the more reasonable when satisfaction of his requirements in their broad terms is at issue. He may in any situation choose between the discrepant items if this offers a practicable solution. Whatever happens there is no extra cost to the employer, but no saving either, as he accepts the solution as satisfying his original requirements. What the employer may *not* do here is to force some solution of his own on the contractor. This may be done only by instructing and paying for a change under clause 12, and then subject to the proviso of clause 12.2.

In any case of discovery of a divergence or discrepancy under clauses 2.3 or 2.4, the party finding it must notify the other 'immediately' in the interests of efficient working. If the employer's requirements or the contractor's proposals are at variance with the other contract documents under clause 2.2 or with one another, when the third recital of the articles applies, neither party is required to tell the other. The employer is more likely to wish to, if only to negotiate a modification to satisfy himself. In the interests of harmony and ultimate satisfaction, the contractor should also disclose his findings.

In summary, the precedence order of documents (in descending order) under clause 2 and the recitals is as follows:

(a) The articles, conditions and appendices, taken together and on the assumption that neither they, nor any insertions into them, are in conflict.
(b) The contractor's proposals taken together, and with any internal conflicts resolved in the employer's favour.
(c) The employer's requirements (including changes) treated likewise, but in the contractor's favour.
(d) The definition of the site boundary, standing in uncertain relationship to the employer's requirements, with any conflict resolved as a change by the employer.

JCT clause 2.5(1.8) – Contractor's design warranty

The question of liability for design under this clause has already been taken as part of the larger discussion of liability. It remains to outline the provisions of the clause as a whole, which is titled as above, although it is not cast in the form of a warranty.

Clause 2.5.1 firstly defines the design, which divides into three sections:

(a) Design performed pre-contractually and embodied in the contractor's provisions.
(b) Design to be performed post-contractually to complete the work under (a) and so satisfy the employer's requirements.
(c) Design caused by a change in the employer's requirements and so modifying work in (a) or (b), or both.

Over all three, the contractor has the same responsibility as given in clause 2.1, and so here the same possible liability. The earlier discussion may be summarised as showing this liability to be that which an independently engaged consultant designer would have towards the employer over a failure to exercise normal skill and care, although this liability may be extended to cover strict care over parts of the design, if this is to be inferred from the employer's requirements.

Clause 2.5.2 defines the contractor's liability as including liability under the Defective Premises Act 1972, when provision of one or more dwellings forms part of the contract. The Act applies to all dwellings and imposes obligations over fitness for habitation (as the primary *purpose*) and other like matters, both upon the contractor as performer of the work and vendor of the product, and upon the employer should he sell later. The provision thus rehearses the obligation as to quality which the contractor already has under statute, and so acts as a reminder that the liability under clause 2.5.1 is extended. It goes on to introduce 'section 2(1) of the Act', which applies when there is an 'approved scheme' for its purpose, imposing a higher level of obligation upon the contractor than the given minima of the Act. At the time of writing there is no such scheme available. This section may be introduced by the employer's requirements specifying a scheme, which is also to be entered in appendix 1 of the contract. The parties both have to act to ensure that there will be 'a document or documents . . . duly issued' for the purpose of section 2(1)(a) which refers to a 'document of a type approved'.

Clause 2.5.3 deals with the alternative situation in which provision of dwellings does *not* form part of the contract. Here the Act still applies, but in its sections 3 and 4. The crucial distinction over liability is that under section 2 it embraces fitness for purpose and cannot be made subject to a limiting amount, whereas under sections 3 and 4 there is no statutory bar on limiting the scope or amount. Accordingly, clause 2.5.3 retains the scope of liability as that in clause 2.5.1 and provides that it is 'limited to the amount, if any, named in Appendix 1'. The clause makes it clear that the amount is in respect of 'loss of use, loss of profit or *other consequential* loss' (emphasis added), so that there is no suggestion of the amount inserted in any way trimming the scope. It excludes from the limitation of amount one aspect of performance, time, by providing that damages for delay are not involved. Even delay due to late *design* is not in view here, as it is not a 'defect or insufficiency' in the design. Equally another aspect of performance, quality of materials and workmanship, does not fall within the clause and liability is thereby not restricted by it.

The term 'limited to the amount, if any, named' could be misconstrued. If no amount is named, this could be interpreted as 'there is no limit', or as 'there is no liability'. If the former is required, for clarity an entry to that effect should be made, rather than a blank be left. No employer is likely to want the latter. Whatever is entered, no obligation is placed upon the contractor to insure, although reasonably he will do so. This aspect was discussed in the closing section on the contractor's liability earlier in this chapter, emphasising its importance and intricacy.

The effect of clause 2.5.4 on the contractor's design liability and its relationship to sub-letting under clause 18.2.2 have also been already covered.

JCT clause 6(5.1) – Statutory obligations etc

While the contract deals in the earlier parts of clause 2 with its potential domestic inconsistencies, it remains subject to the higher authority of statutory obligations, from which no contracting parties are free. The present clause is in the same general mould as other JCT versions, but with important differences due to the contractor's design responsibility.

Statutory requirements

The main thrust is to make the contractor responsible for complying with what the clause refers to generically as 'the Statutory Requirements', and for giving notices and paying fees or charges. These matters are covered in clauses 6.1.1.2, 6.1.2 and 6.2. They are qualified by clause 6.1.1.1 which places with the employer responsibility for any statements in his requirements that any part of them already comply with statutory requirements, that is that he has obtained consents etc. This is most likely to apply over planning matters, discussed further below. (See also the note on clause 6.3.3.)

This initial position works out more tidily than under other JCT contracts, because the contractor is responsible for design. He is therefore not in the link position between the architect and the various statutory authorities under those contracts, seeking to implement the design of the one and not to infringe upon the requirements of the others, although with planning matters he may be commencing where the employer has left off. All the eggs are at that moment in his basket. For this reason, the contractor has to keep the employer supplied with any approvals he receives, as these are not issues to be dealt with by the (missing) architect.

For the same reason, not only is the contractor responsible for paying fees and charges, but the primary provision is for no adjustment to be made to the contract sum for amounts paid, so that the contractor must include for them in the contract sum. This differs from the other JCT forms where adjustment (and therefore addition) is made *unless* an item in the bills or specification specifically requires the contractor to include for particular amounts in the contract sum. Both this contract and the others also allow the possibility of including a provisional sum within the contract sum, approximately covering a defined category of fee or charge, so that the contract sum is not grossly deficient but also so that the contractor is not at risk, as the sum actually paid will be substituted for the provisional one. Where the liability can be assessed reasonably in advance, because it relates to quite normal requirements, it is better for the contractor to increase his own non-adjustable amount within the contract sum. Indeed, only if the employer includes a provisional sum within his own requirements can an adjustment arise, as such sums do not occur in the contractor's proposals under the contract scheme. If therefore the contractor is faced with prospective payments which he cannot assess closely enough and for which the employer has not given a sum, the contractor must secure a pre-contract insertion in the employer's requirements for it to be effective. Otherwise he will be held to have included for the cost.

The other major feature of the present clause is the inclusion within the range of statutory requirements of development control requirements, that is planning matters. Under other contracting patterns, these are the responsibility and the risk of the employer. Here the employer may have obtained outline permission, perhaps subject to conditions. In the extreme he may have done nothing. The contractor takes on from the position given in the employer's requirements, following from clause 6.1.1.1, satisfies himself on it and deals with later arrangements as his design develops. There is thus no provision in clause 6.1.1.2 to relieve the contractor of responsibility or risk for any matter for which he has obtained permission at the date of tender (termed the 'Base Date') or which he still has to settle finally with the authorities thereafter. Any additional cost is, by the silence of the clause, part of his commitment under the contractual 'package' – as is any saving. There are four qualifications of this position in other parts of the clause, and three of them relate to all statutory requirements.

Qualifications

There is, firstly but again, the question of divergence: here in clause 6.1.2 between the statutory requirements on the one hand and the employer's requirements and contractor's proposals on the other, which are classed together since one is the basis of the other. The contractor must comply (as clause 6.1.1.2 affirms) with the statutory requirements; this clause leaves him meeting the cost of introducing any amendment needed for compliance. He also has to secure the employer's consent to his proposed amendment. This may be meaningful over planning matters in particular if there is room for negotiation with the authorities, or over a detailed stipulation in the employer's requirements, although it may be a formality over many aspects of the Building Regulations.

The employer's consent is not to be unreasonably delayed, otherwise the contractor may have cause for extension of time, extra payment or determination under clauses 25, 26 or 28. Equally it is not to be unreasonably withheld. If the employer continues to refuse consent when there is no sensible alternative, determination will again supervene. If there is a practicable alternative upon which the employer insists, the contractor must take it, subject to his rights under clause 12.2 over what must be construed as a change. As usual, 'unreasonably' signals a right to arbitration, but one that cannot be pursued until after practical completion (see under clause 39 in Ch. 10). If the employer's insistence does amount to unreasonable withholding of consent to the contractor's proposal and so introduces a change, he will have to meet the extra costs flowing from it, in terms of design, construction and time.

In respect of divergences, the employer's requirements here also include changes (which are post-contractual) as in clause 2.4.1, over which problems are discussed. In the present instance, changes introduce no special considerations in to the previous discussion about *procedures*. However, as changes usually lead to an *adjustment* of the contract sum, 'entirely at his [the contractor's] own cost' is difficult to apply. The contractor has not had the opportunity to consider the effect of changes in his proposals, as of employer's requirements proper, so that the provision is unfair and could even be misused by the employer. But further the amount of adjustment for any change is agreed *after* it is instructed (or at best at about the same time), so that the contractor may well be able to take account of the correct requirements when agreeing the amount. It would be nonsensical to expect him to agree to part of the cost and not the rest, whether he had noticed the divergence before doing the work or not. All logic requires that he be paid for the correct work and the clause should be interpreted accordingly. Hopefully the courts would agree.

The second qualification over statutory requirements coming in clause 6.1.3 deals with limited work to secure compliance in an emergency. This does not relate to a divergence, but to work already implicit in the contract, such as a matter of health and safety. The contractor has to inform the employer although no adjustment of the contract sum follows. The provision is culled with some amendment from the other forms where there is adjustment, and appears somewhat superfluous.

Thirdly, under clause 6.3.1 a change in the statutory requirements themselves, amending the contractor's proposals after the date of tender, is treated as a deemed change, so that there is an adjustment of payment.

Fourthly, clause 6.3.2 relates solely to development control requirements when 'permission or approval' over some facet may not have been given in full at the date of tender. When given, it may not be what the contractor has embodied in his proposals. This is neither a discrepancy, nor a change in statutory requirements themselves. It will lead to an adjustment of the contract sum just as if the requirements had been changed, unless the employer's requirements say otherwise. Thus they may say in effect: 'Here is the outline planning permission which I have received. You are responsible for obtaining the rest now and later, but either way you will receive the same payment.' The fairness of such a statement will depend upon the issue involved and the risk philosophy of the parties.

Lastly, clause 6.3.3 refers back to clause 6.1.1.1 over any statutory requirements which the employer has stated have been satisfied. Should an amendment to the employer's requirements be necessary due to these limited areas, the employer is to issue a change instruction with the usual effects following. The provision is general enough to cover any mistatement by the employer or a change in the underlying statutory matters.

JCT clause 9(1.11) – Copyright, royalties and patent rights

The purpose of this clause is to deal with costs and to afford an indemnity from the contractor to the employer over any infringement in the areas indicated by the title.

Clause 9.1 is the same in most respects as in other JCT forms. It provides that the contract sum is deemed to include for 'sums payable' over articles etc used 'in carrying out the Works'. There is no reference here, as in other forms, to the works being as described in the contract documents: this is unnecessary as the contractor generally designs and specifies. The contractor is responsible whether or not he mentions any patent etc in the contractor's proposals. He is also responsible for anything arising out of the employer's requirements, again whether they are explicit or not, and should therefore check these carefully when tendering. This responsibility is applicable just as widely to the indemnity provided.

The responsibility under this contract is increased by the addition of copyright, discussed in principle and effect earlier in this chapter. It covers 'drawings, or models of buildings the subject of copyright', which in essence are all such produced by third parties. It excludes those provided by the employer, and so embodied in the employer's requirements. A distinction may thus be noted: if the employer embodies someone else's drawing he must clear the copyright question, if he requires the use of a patented article he takes no action. It would be helpful if the clause referred to *all* copyright materials: there may be written documents, other designed work and so on.

Clause 9.2 is exactly as in other forms and deals with articles subject to patent etc supplied as the result of employer's instructions. The contractor is not liable for any infringement here and so gives no indemnity, while the amounts for royalties, damages etc are all to be added to the contract sum. As the contractor cannot be expected to read the employer's mind, this is generally commendable.

The same two potential flaws exist as for other contracts. While the employer may require a patented article etc expressly, the contractor may choose (as is his prerogative in this contract) to use one as part of *his* design solution to meet an employer's instruction, possibly again without this being apparent to the employer. Whether it is apparent or not, the contractor may provoke legal action over infringement and 'shall not be liable'. This goes too far when the contractor is designing and happens to be negligent over permissions or payments. The second flaw arises when the employer's instruction results in more of something which is already in the contract sum. If the amount of, say, royalties is added to the contract sum and the change is valued under clause 12.5.1 in a way 'consistent with the values . . . in the Contract Sum Analysis', there will be a double payment of royalties. In both of these instances, a reasonable reinterpretation would be likely to have judicial support.

There is no reference to copyright in clause 9.2. A problem here is perhaps less likely post-contractually, but the omission is still unfortunate.

Chapter 6

Tenders, assessment and acceptance

This chapter looks first at the basic issues of risk and of offer and acceptance as operating in design and build, before passing to several issues which are subsidiary in principle, however important in practice, and the related JCT clauses.

Risk apportionment

It is convenient at this stage to summarise a number of issues, already mentioned and yet to come, over which any form of contract has to allocate which of the parties carries the inherent risks. It is possible to place virtually all risks with the contractor or virtually none, with the result that his tender may be particularly high or low. It is usual to steer a path somewhere between the extremes and JCT contracts do this, like others, while maintaining their modified entire contract basis (see Ch. 1). They have been subjected to a volume of criticism from one interest group or another for being too far right or left of centre. For instance, the BPF and GC/Works/1 contracts discussed variously in Chapters 13 and 15 place more risk with the contractor and the likelihood of a higher tender (if not necessarily a higher final account and other costs) with the employer, and this is partly a practical expression of such criticism.

The JCT form with contractor's design passes more risk than the JCT norm to the contractor over design and design-related issues. In other ways it holds closely to the general JCT position, as the following summary of key issues indicates.

(a) Liability for the employer's actions and inactions is specifically allocated to him over various matters. The contractor's own liability is similar, although often by implication.
(b) The employer is responsible for the adequacy of his briefing to the contrac-

tor, expressed in the employer's requirements. He has to bear the cost of any change in his requirements after the date of tender.

(c) The contractor is responsible for all aspects of design and for complying with statutory requirements. He should satisfy himself over the adequacy of any aspects given in the employer's requirements, unless instructed to take them as given. Only if any of the statutory matters are amended or if permission or approval over a planning matter in particular (unless excluded by the employer's requirements) creates the equivalent of a change in the employer's requirements after the date of tender, is he entitled to reimbursement or required to concede a reduction in payment. Otherwise he is held to the contract sum for what he has designed and what he knows about and has yet to design at the date of tender.

(d) The contractor is responsible for producing all documentation other than the employer's requirements and the certified copies of documents immediately after execution of the contract. These latter include copies of his own documents.

(e) The employer is responsible for checking that the contractor's proposals and subsequent information *appear* to satisfy his requirements, but not for checking that they are otherwise correct or adequate in design, or for approving them.

(f) The contractor is responsible for dealing with all site conditions and observing any obligations or restrictions imposed by the employer. He is responsible for designing and, if needs be, redesigning foundations and other works to meet ground conditions and executing them without adjustment of payment, unless the employer's requirements or contractor's proposals provide differently. To avoid overloaded tenders, there is much to be said for introducing limits to the risk, as do other contracts (see Chs 13, 15 and 16).

(g) The contractor is responsible for organising and performing construction and meeting building performance and quality stipulations, including completing the design to these ends.

(h) The contractor may sub-contract any work, including design, with the consent of the employer, but the employer may not require the contractor to sub-contract work and may not designate any sub-contractor, unless he introduces special stipulations into the employer's requirements, which without great care may run counter to the contract conditions and their spirit and be invalid.

(i) While the contractor is generally responsible for the works and delivering them up at completion, some stipulations limit his liabilities over injury. Some insurances are mandatory upon him or upon the employer, who thereupon accepts some risks over damage to the works.

(j) The contractor is responsible for completing the works not later than the due date, subject to particular reasons for the date being made later. While the employer cannot insist on the contractor completing ahead of the date fixed, he can recover liquidated damages if the contractor is late in completion.

(k) The employer is responsible for paying the contractor on account for work and materials at either periodic intervals or progress stages, but may withhold a stated percentage of the value until completion and half that percentage until defects are cleared.

(l) If one party fails to comply with specified terms of the contract or becomes insolvent, the other may bring the contract to a premature end on defined terms.

(m) The contract sum is fixed, subject to adjustment for defined contingencies, including some of the foregoing and design changes. Adjustment also occurs for fluctuations in the cost of labour, materials and some other inputs of the contractor, due to market and statutory changes.

(n) The contractor is responsible for providing any specified operation manuals at completion, for himself remedying defects within a given period and for the cost of remedial work thereafter, until conclusion of the contract. Thereafter he remains liable for design defects and a limited range of construction defects for the usual limitation period.

Obtaining tenders and acceptance

The term 'tendering' carries varied shades of meaning, relating to what a contractor does by way of preparing a tender and to the wider handling of the tendering process between the parties. The former of these is usually restricted to the process of estimating and then forming a judgement on the level of tender to be based on that process. It is convenient to deal with the wider aspect first and then to look at elements of the estimating process, including here design, leaving the question of commercial judgement aside as beyond discussion in the present setting.

Controlling principles

A design and build contract is like any other contract in these respects: a clear offer followed by acceptance without qualification results in a valid contract. This is so whether the offer be so named, or be termed a tender, bid or even a quotation, if given with its intention to be binding clearly able to be inferred. In building contracts there is usually no special requirement about how the contract is evidenced, it may be by exchange of letters, on a form of tender with acceptance, or by other suitable drafted documentation. The contract comes into being before any form of contract is completed, the terms being known in advance. It may even be formed orally before or without all of its terms and details being on paper, subject to the questions of proof and the possibility of implying any terms needed for business efficacy, perhaps in the courts as the last resort. The standard requirement of English law that there must be 'consideration', that is the promise by each party to exchange something of value as part of the deal, is clearly met in almost all building contracts.

The terms and content of the offer must be clear. Usually the person inviting a tender (the standard term in the industry for offers) has prepared documents or caused them to be prepared for any sizeable piece of building work and the tender is based upon those documents. If then there is ambiguity or discrepancy discovered once the contract has come into being it is likely to be construed against the party who produced the document (the *contra proferentem* principle) and not the party who tendered upon it, having made an assumption as to meaning or not having noticed the point. If, however, the tenderer states that his tender is based upon some modification of what is in the documents, he then stands to have any ambiguity or discrepancy in the modification or in its effect on the original documents construed against him. All of this has to be read in the light of the 'appears to meet' question raised throughout Part 2 of this book. Other contracts, such as the GC/Works/1 and ICE contracts, deal somewhat differently from the JCT contract with clashes between documents and the relevant chapters in Part 3 may be consulted.

Acceptance of the offer must be unqualified to give a contract. A qualified acceptance is a counter-offer open to acceptance by the tenderer, whereupon there is agreement and only then a contract. There may be a series of counter-offers, with the option of acceptance passing from one party to the other in turn.

In the course of tendering upon bills of quantities or drawings and specifications, there may be some qualification by the contractor of the tender, perhaps over some specified method or material or over market price fluctuations. This may be followed by a counter-offer by the employer. Any qualification constitutes a counter-offer which, while it stands, effectively neutralises the offer originally invited as the tender, unless it be expressed as an alternative offer. It may be accepted by the employer or be parried by a counter-offer from him, needing acceptance by the contractor. Usually it should be made clear that such acceptance by the contractor is still preliminary to that full acceptance by the employer which will lead to a contract. The position reached is usually quite clear, because most of the terms have been given in one set of documents at one time and are being modified only marginally.

The design and build situation

In the case of a design and build contract, while the principles are the same, the variability of the procedures means that greater awareness is needed.

In many cases, there will be simple single stage tendering, or a simple two-stage tendering: 'simple' here is used to mean that there is an issue of tendering documents and a return of tenders, without a series of questions and negotiations enveloping the process. A JCT code of selective tendering embodies these options. If only one tender is being invited, there is every advantage in extended dialogue all through. It can be useful when there is more than one tender, although it needs controlling in the interests of tendering parity. In general, this chapter assumes dialogue, so following the approach of Chapter 5 over briefing and allowing more issues to be discussed.

If the JCT system is operated in the most simple procedural manner, firstly the contractor receives a number of documents constituting the employer's requirements, which are quite incomplete as tender documents. Secondly, the employer receives back as the tender, not only the offer to carry out the works for a sum of money, but also the contractor's proposals which interpret the employer's requirements and, with them complete the tender documents. (The subsidiary place of the contract sum analysis is held over until later.) The offer is to perform works which the contractor has designed, containing elements of the employer's requirements, developments of them as invited and also possibly changes, welcome or otherwise. These may be technically complex, as discussed later, but also introduce a strong likelihood of counter-offers and so contractual complexities.

If there is a more gradual build-up of the documents with a common view of the project reached by dialogue, more care is needed. There must be an understanding that the parties are treating as a preliminary to a tender, so that all agreements are tentative and dependent on agreeing the whole picture. Neither party will wish to make an offer unawares and have it prematurely accepted! This is equally true when there is a single discussion or negotiation, or when there is competition, whether this proceeds formally by one stage or two (a second stage more often arises informally). Often there are points agreed to early in the stages of design set out in Chapter 5 and which are intended to be fundamental to what follows, while there are many points which are agreed to on the provisional basis that they may be modified in the light of later decisions. The distinction between these types is not of course absolute, or even necessarily conscious. When there is a definite two-stage system, these matters will be clearly defined and should be set out before the second stage begins, so that the parties are clear as to the basis. Firm discipline is needed to tidy matters at defined stages, which is the virtue of observing the RIBA pattern so far as possible, while *all* definite and agreed points should be swept up into employer's requirements and contractor's proposals or equivalents, so that their inclusion in the tender and their precise status are both established. In effect, some of the offering and counter-offering is done ahead of the tender in these circumstances, so that the tender itself can take account of them, rather than be modified afterwards.

In view of the longer period over which the employer may be treating with several tenderers, there is a greater risk that one or more will withdraw before a tender is accepted. If a tenderer has committed himself to a substantial amount of work, he is of course less likely to withdraw while there is hope. It is best for both employer and tenderers to keep each other advised over the commitments as frankly as is reasonable.

Subsidiary principles

Standard principles about parity in competitive tendering need comment. It is usually axiomatic that, to obtain comparable tenders, all tenders must receive the same information and tender in all respects on the same basis. If, for instance, one tenderer raises a query, all should then be supplied with the answer. When competition is employed in design and build, this principle should be followed as far as possible, so that the same employer's requirements are issued to all tenderers in the first place. If there is intervening discussion before tenders are received, and especially if two-stage tendering occurs, it becomes progressively harder to maintain this even-handedness as design development diverges. Inevitably discussion will be about 'different things', so that some matters may be irrelevant to other tenderers. Nevertheless, comparison must be possible when tenders are assessed; the choice is between inhibiting divergence, and possibly good ideas, and measuring its extent so that allowances can be made in tender assessment. This issue has been raised in Chapter 5.

Flowing from this is the need to ensure that all tenderers respond within the required framework. A proposal to move the main entrance from one end of the building to the centre may be acceptable, whereas one to rearrange the accommodation on three floors rather than two may not. If therefore a tenderer is transgressing, he should be warned, as mentioned in Chapter 5 over design briefing. Work which helps no one is pointless, while a correction of course may still be possible. On the other hand, it may be worthwhile to encourage tenders which give alternatives, so long as the original requirement is covered. This could well happen more often as a tenderer may be very advantageously placed to offer something rather different that proves interesting to the employer. The point recurs under 'Competition' in this chapter.

When acceptance of the tender is about to occur then, as usual, all relevant documents should be referred to in the acceptance with any qualifications of their content or purpose. They must then also be incorporated into the contract as drawn up. They are likely to be more numerous and varied than under other contract arrangements. Under the JCT pattern, there is the form of contract (itself strictly four documents), the employer's requirements, the contractor's proposals and the contract sum analysis. In addition there are likely to be letters, memoranda and minutes of meetings and so on. Some of these may have served to achieve a result, but not be very suitably worded as part of a contract. They may also consist largely of irrelevant material. If so, it may be better to embody their essentials into a consolidating document, or even to introduce amendments into the employer's requirements or contractor's proposals, so giving more directness and clarity.

The contractor's estimating activities

Many matters of strict estimating by the contractor are carried out similarly however a project is being arranged with a lump sum or quantitative background. Thus in design and build the price of a unit quantity of concrete or roofing still needs calculation by recognised methods. The prime distinctions arise over the involvement of design and the possible non-availability of precise, or even any, unit quantities of work, according to the state of design and time pressure. Quantities should be produced whenever possible to act as a defined estimating basis, even though they will not usually form part of the contract sum analysis.

The identification of 'The contractor's dilemma' at the end of Chapter 2 is relevant here: how far to go towards final design, and so possible quantification, when the work is heavy in relation to the likely success in obtaining the contract. In the outline following, the exact effects will depend on the stage of design included in the brief and also on how developed a design is expected in the tender.

Design

Design becomes a major factor in tendering, since it is what the employer sees he will be receiving for the price. It also affects the price, according to how economical a design it is. However, as may be seen from preceding discussion, the contractor is restricted by whatever design and specification the employer has included in his requirements.

In respect of design, the first question for the contractor is how organisationally that design is to be produced, that is whether in-house or by an outside consultant or works sub-contractor. The controlling features of this have been explored in Chapter 3. Here it needs emphasising that the contractor is in control of the total process and must have a project manager to integrate the activities of all concerned to achieve the desired results. These persons include those providing feedback data on past projects and their buildability and economy, those producing overall design for the present project starting from wherever the brief leaves off and those blending these elements together.

Control can be especially difficult when dealing with consultants, sub-contractors or specialised construction departments used to traditional patterns of working. They may take quite reasonable views on the 'ideal' solution to the immediate problem addressed by their own section or organisation, but such that they are sub-optimal for the success of the tender as a whole. This usually means an expensive solution or one which conflicts with other considerations like time or convincing a hardnosed employer about functional aspects. Alternatively it may mean stark resistance to change in itself by a long-standing firm or department or an assumption of leadership by a consultant used to the traditional role, even the existence of the 'we know best' syndrome. Any contractor entering design and build needs to assess this sort of problem when analysing the issues to be faced in employing available resources. With sub-contractors, there may be an unwillingness to provide enough information to the contractor ahead of a strong assurance that a deal will be forthcoming.

The process of design is affected by the degree of awareness of the particular employer of what information he has to provide by way of briefing. In the case of a novice employer, the contractor has to elicit much of this by check-lists and so forth, in effect preparing the scope or even content of the employer's requirements. This depends on the contractor's own awareness and perhaps, in turn, on careful use of consultants. While the contractor should allow them adequate scope to secure the information they need, he should also keep closely in touch with what is happening, to the extent of being present at main briefing sessions as well as receiving copies of design decisions. This aspect received consideration in Chapter 3 as well.

Feasibility

In the early stages of design, an employer without his own consultants may seek feasibility checks from the contractor. This sort of activity merges into an early stage of the competitive assessment (discussed hereunder) if there is more than one tenderer and becomes a difficult balancing act for each participant. An unduly pessimistic (or realistic?) forecast may divert an unduly hopeful employer in favour of a more enticing competitor to the point where the opportunity to proceed is lost. On the other hand, an optimistic forecast may lead to later recriminations and rejection after much more effort has been expended. Even this early, there exists a version of the problem which crops up for the employer during tender assessment: that of discerning whether value for money and total price for different proposals correspond in order. Any feasibility figures should therefore be carefully qualified over what they cover and what they exclude in terms of accommodation, facilities and quality, with an ear to anything which may be gleaned of what other tenderers are saying on the same issues. As ever, it is the first figure which almost always sticks in the client's mind, however much the scheme has blossomed meanwhile.

Tender calculation

Design presents features not only within itself, but also in its interaction with the process of estimating for the scheme. This starts with the evaluation for the contractor's satisfaction, as well as the employer's, of design alternatives which may be thrown up as the scheme develops. It continues with the need to estimate for whatever emerges as the contractor's proposals. Some of this may be definite design permitting quantities to be drawn out and priced, when the process may follow that used for other drawings and specification schemes where the quantities are not legally part of the contract. Some of it, perhaps all, may be very sketchy design still evolving as the tender is prepared. This may be adequately covered as incomplete design by statements by way of 'proposals' to be developed post-contractually, but as price it usually has to be covered firmly in the tender.

The standard arrangements under the JCT contract and various others (but see the ICE contract taken in Ch. 16 for some variation of approach in the actual provisions) do not allow for the contractor including provisional sums or allowances in his proposals. Any question of, for instance, approximate quantities for the remeasurement of sub-structure work would need to be specially agreed before being included in the contract documents.

Such quantities, or even provisional sums in the employer's requirements, are restricted by normal practice to items such as foundations where design in advance is impracticable because of physical unknowns in the situation. If the contractor is required to carry the risk for all such site condition factors within his tender, the extent of site exploration work needed when tendering rises considerably. It may well embrace such elements as protecting or diverting existing drains and services, or even subsoil water.

Approximate allowances are not available within the philosophy of design and build to cover what just has not been designed at the stage of tendering but which could be, given time. Here the contractor has to include what he trusts are adequate allowances against what he has yet to design post-contractually, calculated as best he may and within his fixed contract sum. The existence of a contract sum analysis does not ease this aspect, as the individual amounts within the analysis are also not subject to adjustment as design develops, nor should they be held to restrict the development of the design by constituting a 'design to a price' regime. As is discussed in Chapter 6 and followed up in Chapter 9, the analysis stands in its structure and values for various post-contract purposes, among which are the valuation of changes and calculation of formal fluctuations. It is also commonly used for interim valuations when stage payments are not employed (see Ch. 10).

In designing and then estimating the contractor should allow for all components of design which may reasonably be inferred to produce a workable building – and then some more! Thus, while he cannot read the employer's mind over unexpected and perhaps tentative, long-future possibilities of use, he should examine the intended use of elements to see that he has included for all subsidiary details which are a fair inference from the requirements given. Steel angle protection of concrete arrises in areas where heavy trolleys are to operate is normal practice and needs to be foreseen and included in the tender if the presence of trolleys has been identified.

The aspect of making allowances may need care in eliciting from external consultants what they can foresee. They are not in the position as in other circumstances of being able to raise a variation extra; here their side of the contract has to bear the cost, although it is likely to prove difficult to pin responsibility on them for failing to warn in such areas. Sub-contractors are also awkward here: unless their works are entities, they may well consider that they should price only what is explicit. In so far as they do not have a concept of the scheme as a whole, they may well be justified. The contractor may have to make his own allowances on top of what they quote.

All of the points given above are relative detail, even the matter of substructure design. A more drastic uncertainty may be that of planning permission, if this is left with the contractor. Indeed everything points to the advantages of the employer dealing with this aspect ahead of tendering, and many do, except to the extent that elements of it depend upon the contractor's design development. How the contractor deals with it will be conditioned by the time available: whether he can clear it and price accordingly or whether he has to treat its outcome as one of the risk items in his tender. There is much to be said for the employer's consultants taking it on during tender assessment, subject to time constraints. The JCT contract allows a deemed change to arise post-contractually in given circumstances, leading to an adjustment of the contract sum as discussed in Chapter 7. In general, the contractor is advised to make the position clear to the employer and urge him to deal with planning matters ahead of or otherwise apart from tendering.

An element in the tender is the cost of producing the design itself. This cannot be assessed by any single rule of thumb or percentage, as the extent of employer's design affects how much the contractor has to produce. It is also the case that some design costs are contained within prices quoted by specialist sub-contractors. Beyond this, the contractor has to include the rest in his tender. Further, as with his estimating costs, he should allow appropriately for design costs of projects for which he does not obtain the contracts.

The tender as presented to the employer should conform with any stipulations over format and detail set out in the requirements. In the immediate context, this includes complying with any required structure for the contract sum analysis (see discussion on this document in Ch. 9). If there is insufficient design or time to permit a proper apportionment of the tender to the headings of the analysis, the best possible has to be achieved. When once the analysis is incorporated into the contract, its detailed amounts become significant for several contract purposes, so that 'stab' figures may lead to unwanted problems, financial disadvantage or even advantage post-contractually. It may be wiser to provide a provisional analysis with the tender and then revise its distribution while assessment proceeds and a cooler view is taken. It is not unknown for tenderers to hold back on their analyses until asked for them during assessment, when they can be adequately explained. This may avoid an unaware employer jumping to conclusions by comparing particular elements within the various analyses when there is competition. Variant designs may well mean very varied sums for the 'same' element between schemes.

Time and qualifications

These two aspects tend to go together, the less there is of one the more there will be of the other.

The question of time lies at the heart of many problems faced by tenderers for design and build work. The period allowed is almost always inadequate. Some clients fail to realise entirely that the extra function of design taken on by the contractor needs extra time, in part because it has to proceed end-to-end with estimating and assessing programme implications. This may be so even when they are clients who have commissioned construction by traditional means and so have had an independent design period ahead of that for tendering. It runs through the question of considered design itself, into estimating for the reasons outlined above and on into the related complications of obtaining sub-contract design and prices within the uncertainties. All that can be advised is that the contractor seeks more time than is offered – and then presses on as usual when he does not get it! Perhaps his only greater problem in design and build tendering is that of greater costs in tendering for projects which do not turn into contracts for him.

There will regularly be numerous qualifications and provisos about a design and build tender, as against the employer's requirements, even when tendering has been a two-stage or entirely negotiated process, so that points can be aired on the way through. Further and whichever way matters have been conducted, any oral discussions should be recorded for ultimate incorporation into the contract documentation. As some of the items cropping up during any formal tendering may not lend themselves best to initial explanation in writing, a post-tendering interview is essential so that such points may be aired before being recorded. If there has been competition, it gives the contractor a useful opportunity to explain the virtues of his scheme as against those of his competitors, who may of course view the occasion in similar light!

Competition

There are two primary issues here for the employer: whether to have competition at all and, if so, how to set about it. They also rebound on the contractor heavily. The distinctive features in design and build work are the extra work involved because design is competitive and the problem of assessing competitiveness in this situation.

Value of competition

For the employer, there is a prima facie economic argument for competition, as is always the case. This is more particularly true if he is commissioning just one project, and so is not too concerned about who carries the costs of unsuccessful tendering. If he is regularly in the business of procuring contractor-designed work, he may well decide that it is in his long-term interests not to have competition over any one project, but to secure it by using a panel of contractors on a rota basis and discarding any that fail to give satisfaction. This thins out the costs of tendering which *must* be borne by someone, although not necessarily this employer. Alternatively, the employer may seek competition only or mainly in the first stage of a two-stage procedure and ease costs that way. Sometimes he may have developed a close relationship with a single contractor, so that the element of goodwill is strong and working routines are well established, thus compensating for any marginal extra price level.

There are just a few specialised projects where only one suitable contractor may be available, so that competition is precluded. It may be suggested by contractors over some other projects also. In any case, without competition, the employer must either accept the price he is given (however he may deal over the design) or be able to negotiate. The latter assumes either business muscle or the use of his own consultants, who are best engaged to work with the contractor in building up the price alongside the design.

The value of competition to the employer is not only its economic possibilities, but that it generates different design ideas. So may the traditional design process, but the ideas are usually all from the same source. When competitive, the design and build approach is more like an architectural competition. In the same way, it should be carried only to a first stage before the field is narrowed. There is the need for propriety if ideas from one tenderer are in any way to be injected into another design, and this is discussed in Chapter 5, with the question of copyright.

Cost of competition and selectivity

For the contractor, there is a prima facie economic argument against competition. It is not just the odds against him securing the contract, but the level of expense involved in designing in enough detail to be able to prepare the tender, discussed under 'The contractor's dilemma' in Chapter 2. The strength of the argument varies with the number of tenderers, how early they are introduced in terms of the RIBA stages (see Ch. 5) and how far through these stages the competition runs before a selection is made. The possibility of obtaining basically unsuitable designs should have been guarded against at an earlier stage by developing the design information incorporated in the brief to include all mandatory points and obtaining planning permission, as discussed in Chapter 5, provided that this is not so done as to act against obtaining enough design variety at the present stage.

Although it is not done very often, it is possible to offer a fee to each prospective tenderer, which becomes payable if he is not awarded the contact. Whoever is successful includes all design costs, and so the fee, in the contract sum as usual. With the fee set at a level which only partly reimburses tenderers, this system has the virtue of discouraging the employer from seeking designs frivolously and of discouraging tenderers from running businesses financed by unsuccessful designs. It could well be used more often. But safeguards are needed over the stage that is to be reached and the standard and extent of what is to be produced.

Mention has been made of two-stage competition and logically there may be quite a number of stages, however imprecisely they may be identified. The first comprises a selection of a list of tenderers. Possible tenderers should be interviewed, as they often are, over their experience in the field and their management resources and methods in design and construction. These aspects should be filled out with inspection of projects in hand and completed and interviews with those who commissioned and who are using them. If tenderers intend to use external consultants, the process should be extended to them, even though formal subletting to them has to be cleared post-contractually (see JCT clause 18).

It is a reasonable broad principle that the earlier tenderers are introduced, the earlier it may be possible to eliminate them or, putting it positively, to become more selective. If several tenderers are being treated with at the inception and feasibility stages, it may become clear that some are not suited to the particular project, even if they may be borne in mind for other types in the future. It should be a firm decision to carry out the main selection at a clearly defined stage by a named date, so that everyone knows his position, although in practice affairs often tend to be less precise. After this it may be inequitable to ask more than one tenderer to proceed to finality, except perhaps for off-the-peg buildings. One tenderer should suffice at this stage, provided enough controlling points are tied down in advance, as considered under the next heading. Someone else can always be kept in reserve to give any needed incentive. The less significant the scheme, the less ponderous need the whole tendering process be – although the once and for all employer may not see it like that.

In the absence or exhaustion of competition, it becomes necessary and legitimate to negotiate in a normal way over issues not subject to this commercial spur. Competition may be conducted in the quasi-clinical isolation that usually obtains when all design etc comes from the employer, or it may proceed with a level of interchange between the employer and each of the tenderers, as has been suggested earlier in this chapter. If so, something of a Dutch auction may be engendered, as similar points are aired with the tenderers in turn. There is nothing unethical about this, provided proper standards are maintained. A check on parity for employer and tenderers alike is introduced by a formal intermediate stage, when all dealings are brought to a common level.

Scheme assessment of tenders

Some overarching considerations

Some of the points mentioned over the progress of competition indicate that dialogue with the tenderer or tenderers over *development* of the scheme (a major topic in Ch. 5) also includes an element of *assessment*. This possibility is a distinctive part of design and build and, if not strictly necessary, is very useful when a tenderer is doing so much more. A longer look at his scheme is valuable, while seeing it develop replaces the usual consultation by the employer with his own designers, thus allowing changes to be phased in at appropriately early stages. However, for ease of discussion in this main section this progression is ignored and assessment is viewed as a single stage operation, covering all aspects except finance, which is reserved for separate treatment and which should be assessed separately.

More elements are in play in this type of contract. In others, it is usually price that predominates, assuming that the suitability of tenderers has already been cleared and that the contract period has been given as fixed in the tender documentation, as is common, so that the priorities of competition are limited to price. Any special points that have been asked for as part of the tenders will be subsidiary in character. What is being offered in a design and build scheme, the works themselves, is the major variable in the tenders. Because of this, it can be sensible to allow tenderers to state their own contract periods when possible, as these may be influenced by the type of construction offered. The relative importance of the works, time and price will vary according to the employer and the project, and he should establish his policy on these ahead of receipt of tenders, but be aware of how he is prepared to vary his policy in the light of what he receives.

In view of the several strands within tenders, it is important that assessment has matching strands. If the employer has several consultants, each should be given sufficient time to examine his strand and this time will vary according to the nature of the project: there may be critical aesthetic or planning considerations, special ground conditions or structural needs or the services may be unusually complex or innovative. Any such aspects will need to be set against the more normal ones to establish balance and check how they interact. Scheme assessment of the tenders assumes greater significance as pointing to the best selection when the financial amounts given are relatively close and in themselves little guide. Unfortunately, the great temptation in practice is for the employer to look almost solely at the tender sums and to pay inadequate attention to what he is being offered in return for the cash he will be laying out. Both of these legs of assessment should go on, and preferably be conducted separately and in parallel, before their respective weightings are considered.

It may be necessary to ask tenderers for supplementary detail to explain what they propose, or even to secure what was asked for in the employer's requirements, but which has been left out by accident or in hope that it was not really needed. A short tendering period is particularly likely to lead to this sort of omission, as well as other mistakes. Some tenderers may have considered that details sought about their staffing, programme or method statements were unduly intrusive (sometimes they are!) or restrictive. Their omission may, however, indicate a poorly thought out submission, even though the technical detail may be quite well presented. Any such features should be noted initially and given care as assessment proceeds.

In view of the frequent tendency for offers to vary quite widely in several respects about the employer's intended focus, it is sometimes helpful in the early stages of assessment to pick out one tender, if there is one, which appears superficially to conform most closely to what has been requested as a scheme and as a sum. It may then be possible to relate the other tenders to this benchmark in terms of how far away from the 'norm' they appear to be on various criteria. This process should not replace the more rigorous assessment outlined below, but may instil some sense of order into an initially confused scene. It becomes unwise if it takes over and leads to a determining judgement before greater objectivity is brought to bear.

If there are divergences from the employer's requirements, it may become a matter of commercial expediency whether to insist that they are removed, to ignore them or to take them into consideration. In principle the employer may think that he has asked for precisely what he wants, but in practice one or more tenderers may have produced a proposal which is compellingly better in some way. This leads to a dilemma comparable to that mentioned under 'Unfinished design' in Chapter 5: whether to shop the idea around the other tenderers and, if so, whether this is infringing confidence and possibly copyright. Alternatively, it could be seen that not to allow the tenderers to come in on the matter would be giving an unfair advantage to a tenderer who has not followed the ground rules. Often the matter may be seen as an ingenious development of the requirements, so that the advantage remains with its originator. Certainly, any wholesale passing around of ideas is unethical and also could lead to action for infringement of copyright. If the contractor considers that any part of his proposals could be subject to this sort of treatment he should cover his tender with a statement of confidentiality. There is much to be said for making sure and doing this every time as part of the proposals.

A particular case of this dilemma is that in which all tenders but one are outside the employer's budget and do not lend themselves obviously to an adequate scheme amendment, while the remaining tender meets the budget by trimming the quality. The cheap option could be accepted if suitable, but if the trimming is unacceptable, that tender should drop out for that reason. In either case, the other tenderers could be asked to reconsider their schemes against an indication of the budget position. The conditioning factor here might well be the time question, as substantial redesign may take matters beyond what the employer is prepared to contemplate. All of which points up the virtues of initially having a good cost assessment and a realistic time-scale embracing all stages of the project.

Documentation

Whatever the policy issues just discussed, there are several standard points about reviewing the documents making up the tender. Under the JCT arrangement, these documents are the contractor's proposals, with the contract sum analysis as something of an appendix. The documents must be checked in terms of what is being offered as a scheme, and this is the substance of the next section. It is also necessary to ensure that no qualifications have been included that modify the employer's requirements over any of the proposed terms of the contract, particularly over the apportionment of risks set out at the beginning of this chapter. In many cases, such an attempt would founder on JCT clause 2.2. if this applied, unless it were effectively drafted in at the contract stage. However, it is as well to clear up such points early, with the cost of changing them, rather than face a hitch in entering into the contract or a dispute later.

For similar reasons, the documents must also be checked for completeness: there should be an outline list of what is to be supplied in the employer's requirements and a corresponding list identifying all documents in the contractor's proposals. Any discrepancy should be queried: whether a document is missing from the bundle or from the list, its standing must be clear as it is due to become a contract document.

Completeness extends to the stage that the design has reached, which should correspond in principle to that stipulated in the employer's requirements. There should be adequate delineation of how, in method and content, the balance of the design will be completed. In particular the state of planning permission, if not already fully obtained by the employer, should be acceptable.

The contract sum analysis should contain no provisional sums, other than those included in the employer's requirements. Everything that is given in the contractor's proposals must be tendered for on a firm basis, so that the contractor will carry any adjustment in his costs for elements of the design which he has yet to

on should be given to the interface between the design as given nployer's provisional sums: what is *actually* in the former may *intended* in the latter. In addition, it should be checked that nent has not invalidated the principles laid down in the em-uirements for pricing preliminaries costs associated with provisional sums as being for defined or undefined work, or whatever method of designation has been adopted. These principles are set down later in this chapter under 'Contract sum analysis, structure of'.

The case against including measured quantities in the contract sum analysis and other points over them are introduced later in this chapter and taken more fully in Chapter 9. If any are included and they are to form part of the contract, their precise status in this respect must be clarified. If they are given as firm, they should either be checked if there is time, or so qualified that the contractor carries the effect of errors, a point aired in Chapter 9. If they are given as approximate, their broad accuracy as a reflection of the employer's likely commitment should be checked. So too should the rules for controlling the design and measurement of the work represented, because they are like provisional sums in their effect and the amount payable by the employer is variable.

Satisfaction of employer's aims

This aspect of assessment passes from formalities to the substance of the project. It deals with how some of the employer's aims, considered in more detail in Chapter 1, are met by the contractor's proposals. The value aspect, which all the present factors affect, is reserved for the next section.

Major features

It is a fairly objective matter to assess the functional aspects of the scheme at three levels of detail:

(a) The extent of accommodation provided in total area, units, heights or whatever is appropriate is likely to have been set out in the employer's requirements, but should be checked. This may be subdivided into habitable or other usable space, against circulation, services and other non-usable space, in terms of the purpose of the scheme. Other standard measures, such as car parking spaces, may be applied to external facilities. Within the same brief, a tenderer may respond with very tight accommodation in some ways, with reasonable adequacy or with excessive provisions in places. While the last may be an aspect of what is still a cheap scheme as tendered, it may mean for instance higher life costs or suggest ways to initial economy.

(b) The efficiency of the layout so far as left to the tenderer should be reviewed. This is partly a question of total space as under (a), but involves the relationship of rooms to each other and to accesses, room shapes, orientation, daylighting and all the regular ingredients of planning. Again external facilities and the interrelation of buildings, if more than one, need consideration.

(c) Technical aspects of the structure, furnishings, fitments, services etc should be sifted closely to ensure that nothing is below the minimum standard acceptable for the scheme, in construction or in functioning. This may be a matter of checking the adequacy of statements of intent, where design is incomplete (but see comments below).

It is anything but an objective matter to assess the aesthetic and related qualities of the proposals, externally and internally, but it must be done, for even the most lay of employers will have a view (and probably twelve if 'he' is a committee of twelve), to which he is entitled. After all, he is paying for the project and will have to live with it! Any consultants of either party should be listened to carefully by him here: this is an area in which oral discussion and visual demonstration are most valuable. Certainly, the scheme should not just be taken for granted. Beyond this, the present work cannot go within its terms of reference.

The proposed contract period is very often stated in the employer's requirements and, if so, should be checked whether it has been accepted by the tenderer. If it was not stated, the tenderer's period should be reviewed as to its feasibility, neither over-optimistic nor unacceptably long. A method statement may have been asked for in the employer's requirements for a complex or novel project. Either that or a master programme (not mentioned in the present JCT form) is useful for this part of the assessment. A reliable period is needed, because both parties will be bound by whatever is agreed, at pain of extensions of time or liquidated damages (see Ch. 8). The financial effect of the period is looked at under the next section.

A major caution

In the whole of this aspect of assessment, function, aesthetics, quality and time, it is most important to bear in mind the statement in the third recital of the JCT articles of agreement that 'the Employer has examined the Contractor's Proposals and . . . is satisfied that they appear to meet the Employer's Requirements'. Even under other design and build forms, the same principle applies. This is discussed in several other places, especially in Chapter 5 over design responsibility: the principle being that the employer should not give approval to any proposal in such a way as to relieve the contractor of responsibility, by suggesting that it has been checked and found adequate. The employer, and his consultants who may be tempted further, should maintain the 'appear to meet' stance. Even if something is challenged as possibly or definitely inadequate, the suggestion of an alternative should not be couched in terms of 'this will definitely be acceptable', but merely 'how about this possibility?'.

What is being assessed may be two types of material: proposals which are really statements of intent over work still to be designed, and firm proposals backed by precisely drawn or written information. The former type, by its very nature, is far more likely to fall clearly into the 'appear to meet' category. Exceptions are such statements as 'the undesigned tank room will have a floor area of 40 square metres' or 'the heating system will be capable of providing a temperature difference of 25 °C between internal and external air'. The employer could not then come back in the post-contract phase and say that these statements only appeared to meet his requirements, which were for 60 square metres or 30 °C respectively.

With the latter type of material, hopefully the majority for most schemes, it is largely the case that the employer must be held to know whether the proposals actually meet his requirements. Thus matters of space, layout, appearance and time are there for him to see, whether appreciation of their quality is judged to be objective or subjective. It makes no defence if he should later argue that he could not read the tender drawings: they showed factual matters and he should have found someone to explain them if needs be. Only over technical aspects, such as those under (c) above, can he claim that appearance has not been borne out by reality, as when something fails structurally. Even in technical matters, he must reasonably be held to have understood what was meant by 'vinyl flooring', even though he may not be expected to understand differences in composition or grading. On the *contra proferentem* principle, he is entitled to a quality 'reasonable' for his project and even to choose between tiles and sheet if this is not specified.

It has been suggested under (c) that technical standards should be checked to ensure that they are adequate. This is to try to eliminate the undesirable at an early stage, although possibly leading to an upward adjustment of the tender. But if the employer is satisfied with what he finds or with what is put in its place, this is purely for his own peace of mind and not to be stated formally to the contractor. Two individually satisfactory proposals may combine unsatisfactorily, and the employer should reserve his position against such an eventuality. It is the scheme as a whole which he ultimately accepts.

Financial assessment of tenders

Even if the contractor's proposals fit the employer's requirements like a glove in all other respects, there still remains the question of the price. The presence of competition may not give an easy answer, because the ranking order of schemes and their prices may not coincide.

It must be faced that there is no magic formula that readily combines all the considerations, so that amounts can be extracted from the contractor's proposals and dropped into the spaces to give a direct answer. Nevertheless, something like this must be done in principle, however different it may look in practice, if several schemes have to be compared. If there is only one, it must be set against some notional standard. Before taking the idea of comparison further, the main sets of considerations for a scheme may be reviewed.

Value considerations

These have been described under 'Satisfaction of employer's aims' above and those relating to the finished product may be summarised as:

(a) Quantity of accommodation.
(b) Efficiency of layout.
(c) Technical quality.
(d) Aesthetic quality.

In general, the higher the score on any one of these the better, although there might be an over-provision of quantity in (a) that detracts from value. Layout in (b) has its effect on quality in (a), while its value here is partly that it saves running costs as considered below. Double counting has therefore to be avoided. Of quality in (d), only the difficulty of assessing it need be noticed again.

The interrelation of some of these considerations and of those under the next heading is illustrated in basic terms in Table 2.1.

Capital cost considerations

The major element here is the tender figure, which is more easily considered if the contract sum analysis is available. The detail of the analysis itself is discussed later in this chapter, and it may be seen that the JCT recommended structure for it is somewhat different from the RICS recommended cost analysis, used in relation to cost planning. Some reconciliation may therefore be needed.

The analysis may vary in structure between tenderers and also in the relative weightings of sums within it. Variation may be due to the amount of discretion allowed to tenderers in designing their schemes or it may be due to some of them choosing not to follow the analysis required by the employer. Either way, more divergence between tenderers' analyses is likely than occurs between priced bills of quantities (and that can be noticeable), so that comparison of competitiveness based on the details of analyses will be limited. Any distortion that looks excessive should be explored, if the tender concerned is being seriously considered. If there appears to be such a distortion, the ultimately favoured tenderer should be asked to review his analysis before it is incorporated in the contract, although as with bills of quantities strictly he cannot be forced to change what he has given. Often an analysis as put forward will also lack detail and it may be necessary to amplify it before incorporating it.

For present purposes the quantity and quality of what is being provided and already assessed must be taken as given, as would happen in assessing a tender for a project designed by an independent consultant. The question therefore is whether the tendered sum is reasonable for the scheme as designed. It may be answered by using such of the standard techniques as may be suitable. If there is a cost plan already prepared by the employer or his advisers, and this is for a scheme proposal relating closely enough to that now being considered, this may be used as a basis for comparison. If not, some form of cost plan or quantification will be necessary. Whatever is done, allowance must be made for the following:

(a) The contractor's consultant or in-house design and related costs.

(b) The inclusion of any contingency for risks due to design not settled.
(c) The effect of integrating design and construction.

These do not all affect the cost in the same direction. In the case of (c), part of the effect already shows in the type of work resulting from the design, and so in its quantity. 'Buildability' rather than quantity, and its effect on the contractor's programme, are what is now in view.

A forecast should be made of the likely final account for the project, in the light of its specific design. While the contractor will be responsible for producing the scheme as designed or as yet to be designed, in the main, there are elements that are open to financial adjustment:

(a) *Provisional sums*. Although these are as included by the employer, the nature of the contractor's proposals may affect how much will be expended and so included in their place.
(b) *Provisional measured work*. The same consideration applies over elements like foundations if these are subject to remeasurement, and so not at the contractor's risk, whichever party is going to determine their actual extent.
(c) *Changes or variations*. The employer may be considering instructing these in due course and the contractor's proposals may be more or less suitable for accommodating them easily and economically.
(d) *Fluctuations in market costs*. Particular items only are allowable here and, according to the design and the contract time span, even the same items may have varying effects on the final account. If clause 37 applies, appropriate lists of basic prices should be included as described thereunder.

It is possible that examination of the figures in the contract sum analysis may reveal an error of arithmetic. All being equal, the normal principle should be applied of asking the tenderer whether he will stand by his total, while having the figures adjusted within the total to remove the error. The usual routines for this should be followed, so far as the form of the contract sum analysis lends itself to them. If the tenderer is unwilling to stand by his total, he must be asked either to withdraw or what amendment he requires. Which option is used should be decided in advance of the necessity arising. In view of the differences in schemes, the possibility of amendment is likely to be attractive, and, if there is only one tender, it becomes inevitable.

The next most important consideration to that of the tender figure is the contract duration. This is reflected for the contractor in the tender price. For the employer, the comparative *value* of various durations in terms of obtaining use of the project has already been discussed. The *cost* refers to the time for which his money is tied up and earning him nothing. This should be evaluated by standard discounting techniques.

A marginal consideration that may occasionally be relevant is that of the fees of the employer's consultants. In the extreme case of an unusual design, the extent of involvement of these consultants may vary and any fees applicable with it.

Continuing cost considerations

These are in no way peculiar to design and build contracts and their relative importance should have been given in the employer's requirements. Because the contractor is in competition (potential or actual) over design and price, there is the peculiarity that he may have been tempted to design with just initial costs in view rather than total costs, thus leading to higher continuing costs. Again, discounting embodied in cost-in-use techniques should be used to strike a balance among these continuing costs and between them and capital costs. The more important elements have been referred to in several contexts in different chapters and in summary outline are:

(a) Maintenance and cleaning, regular and occasional.
(b) Energy costs.
(c) Flexibility and need for alterations.
(d) Useful life of building.
(e) Disposal costs, for demolition or otherwise.

In addition to taking account of the whole costs of all these, it is necessary to allow for the differential costs of staffing the project, between it and its competitors. If a building needs two extra receptionists or if three more staff are needed because of walking time in longer corridors, these features should be considered.

If the project is to be rented or sold, at once or later, the question of return will have entered into the deliberations over whether to go ahead at all. In these the value of the land will have been included. The efficiency with which the project relates to the site should be reviewed to see whether any of the calculations should be amended.

Comparison of value and cost

This heading is an alternative way of saying 'value for money' and brings together the considerations given under the preceding headings. The most difficult part of tender assessment, and the one requiring the most experience, is the evaluation of the various considerations. He is a brave employer who tackles it alone on any but the most simple of schemes. He may well need advice from consultants of distinct expertise, especially between 'value' and 'money'. These various pieces of advice have then to be combined into a decision. In what follows a somewhat theoretical model is used to describe the steps that should be taken.

In principle, the money side of the balance is the easier. Just as the various elements of capital costs may be brought to one total and similarly those for continuing costs, so the two may be added together to give total cost in the common denominator of money discounted to present value.

On the value side of the balance there are numerous elements differing in nature which, for tidiness of discussion, have been grouped under four sets of value considerations, although the employer may need to split one or more to refine comparisons in practice. According to the employer and the project, the relative importance of these considerations will vary. Furthermore, there is only limited scope for matching them separately against elements on the money side. Quantity of accommodation and technical quality tend to vary directly with capital cost; while efficiency of layout and technical quality tend to vary inversely with continuing cost. Aesthetic quality is more elusive.

When this limited matching has been done, it is useful to arrive at a global value. This may be done to compare schemes, or the one being examined with a notional scheme, by giving value weighting to each set of considerations, according to relative importance to the employer. Quantity of accommodation, which is basic, should be weighted first, with the others treated as adjusting factors. If this weighting is performed before schemes are compared, at least a measure of objectivity is introduced. The various sets may then be given comparative scores across the schemes and weighted value totals calculated.

By this rather crude method some attempt to quantify subjective judgements along with the rest may be made. No doubt there will be loose ends, and the result may be adjusted with hindsight over the weightings. If a clear order does not emerge from the money and value rankings, they may be brought together by calculating cost per unit of value, and these results may be ranked.

With all this illusory science performed, the employer still has to decide, if he has a choice, which is demonstrated as the last step in the following narrative. This relates to Table 6.1 which shows widely different tenders to highlight the illustration:

(a) Tenders are received and adjusted for different construction times, giving capital costs: there is a widespread suggestion that 'A' is of low standard, 'B' average and 'C' high.
(b) Continuing costs are calculated at present value: these are much closer together, with 'C' coming marginally below 'B' on these figures.
(c) Total costs remain in the same order: the total spread has increased in amount, but is down in proportion, while 'B' has moved up in relation to 'A' and 'C'.
(d) Weighted values on quantity of accommodation give 'B' a small advantage over 'A' and 'C'.
(e) Efficiency of layout favours 'A'.
(f) Technical quality swings heavily towards 'B' and 'C'.
(g) Aesthetic quality accentuates this trend, with 'C' gaining quite heavily.
(h) Total units of value are in the same order as total costs.
(i) Capital costs per unit of value are fairly close, with 'B' beating 'C', and 'A' least favourable, reflecting tender levels and scheme quality.
(j) Continuing costs per unit of value give 'C' best, followed by 'B' and then 'A', with a wider spread than on capital costs per unit and reflecting scheme quality.
(k) Total costs per unit show the same order, making 'C' the best scheme as value for money.

Table 6.1 Comparison of tenders illustrated by a quantitive theoretical model

Costs		*Tenderer 'A'*	*Tenderer 'B'*	*Tenderer 'C'*
(a)	Capital costs	352,000	426,000	540,000
(b)	Continuing costs at present value	286,000	335,000	324,000
(c)	Total costs	£638,000	£761,000	£864,000
Weighted values				
(d)	Quantity	2,000	2,100	2,000
(e)	Efficiency	+10%	—	—
(f)	Technical	—	+20%	+30%
(g)	Aesthetic	—	+25%	+50%
(h)	Total units	2,200	3,045	3,600
Cost per unit of value				
(i)	Capital	160	140	150
(j)	Continuing	130	110	90
(k)	Total cost per unit	£290	£250	£240

The significance of these figures is discussed in the text.

The employer has as many facts and statistics as ingenuity can provide, and now must decide:

(a) On actual costs, 'A' leads on capital, continuing and so total costs, but is down heavily on value. If it satisfies the employer's requirements at least at the threshold it may nevertheless be worth considering. This might apply with a minimum cost industrial building, although this employer appears to have weighted quality heavily.
(b) On costs per unit of value, 'C' leads, although 'B' is close. The employer might prefer 'C', but hesitate for two reasons. One is that 'C' gains on continuing costs only, and these are the more speculative aspects – either way. The other is the 'banger versus limousine' dilemma or, more formally, capital rationing: it is better, if you can afford it. Just as he has a threshold of standard, so he has a ceiling of price.
(c) In between on most figures lies 'B'. It has the highest continuing costs by a small margin, the highest weighting for quantity of accommodation and the lowest capital cost per unit of value. It could thus be a serious contender,

especially if the employer decided on review that he had weighted technical and aesthetic qualities too high for realism. A quarter off the percentages would bring the unit of value marginally in favour of 'B' and the cost per unit decidedly so. 'B' and 'C' costs per unit would straddle that for 'A'.

A criticism of this example is that placing the figures widely apart for illustrative purposes produces exaggerated differences, with exaggerated swings if assumptions are changed. This is not impossible in real life, but would suggest either very loose briefing that failed to define priorities adequately, or that at least one tenderer ('A' or 'C') has radically misinterpreted requirements. It is also quite possible in reality to obtain very close tenders in this type of work. A number of further comments may be made about this example:

(a) It underlines that like is not being compared with like.
(b) It indicates the value of a progressive or staged assessment; this allows the employer to weed out undesirable interpretations before they get so far, and may allow a tenderer to change direction early to his and the employer's benefit.
(c) It shows how subjective elements enter into the assessment as they do in briefing the traditional independent consultant, where they are masked because only one design solution emerges.
(d) It contains the same elements as in other assessments, but they occur at a different stage or stages of the process.
(e) It is a somewhat artificial or stylised model, but the thought processes that are in it in embryo have to be gone through in practice in some form to give as balanced an assessment as possible.

Awarding the contract

When assessment and any negotiation are complete, a tender may be accepted, following normal procedures with some variation of emphasis. The contract documents should now be prepared, incorporating all the papers referred to in the preceding sections. They are likely to be extensive, because of the variety of elements needed, such as planning permission, and not included in other contracts.

A pre-contract meeting immediately before acceptance is desirable. This can be used to run over points discussed during assessment to ensure that they are properly taken into the contract documents and in general that the documents are comprehensive. Those acting for each party should be introduced, their functions and delegated authorities cleared, any sub-lettings should be proposed and aired and routines for communications established. If these items are not actual contract material, they should at least be incorporated into fulsome minutes.

The programme and procedure for completing the design and obtaining statutory approvals are particularly important. The responsibility here is entirely with the contractor, but the employer and any consultants will need to be aware of when they have to act in any way, or even when they may receive data for comment. It is not a bad thing to emphasise and repeat (as this book does) that the employer does not assume responsibility for design by virtue of receiving it, or commenting upon it. According to the extent of design still to be completed, the period between awarding the contract and commencement of work on site may vary quite considerably. The employer may wish to press for an early start, but if he has already gained time by obtaining tenders based on very limited design, he should in fairness allow enough margin now. This is, however, what very often is *not* done; sometimes less time is allowed than for a traditional project. Whatever the state of design, the contractor is entering into a firm commitment on price. He therefore needs sufficient time to develop his design to avoid undue risk of loss on this score. The less design there already is, and assuming that the contractor has not skimped it so far, the more time reasonably should be given. If he has planned ahead, he may well have stipulated this when tendering.

Contractual insurances by either party should be reviewed to see that they are adequate, while insurances and terms should be approved. Design insurance is not required by the JCT contract, but any stipulations in the employer's requirements should be checked over, including any limitations on levels of liability.

This is the last chance for the employer to check before commitment that he is able to provide all that he must in time. He should by now have passed all planning questions over to the contractor, but he still has to provide the site and any special accesses on time. Soon, perhaps before handing over the site, he may need to provide instructions about expending provisional sums, and give information about any materials or equipment that he is supplying, or direct contracts that he is organising. All these will be described in the employer's requirements, but have a habit of needing a lot of work all of a sudden. The employer's other big provision is finance: he should review his budget and cash flow. The contractor is usually only too pleased to provide expenditure forecasts, to which the employer may add for his own consultants and other such amounts.

Contract sum analysis

It has been possible to discuss tender assessment without direct consideration of the contract sum analysis, because this is a subsidiary document to the contractor's proposals, although still a contract document under the JCT form at least. It should be obtained with or soon after the rest of the tender, as it is very useful in performing the functions that have been discussed, while it needs to be agreed in detail as being suitable for post-contract use.

Purpose of the analysis

A breakdown of any large sum of money usually helps in spending it sensibly. This is true of the contract sum in a design and build contract, even though the contract is at law basically an 'entire contract'. This means that it is in essence an indivisible commitment by each party to perform his side of the bargain in its completeness, either to design and build the works or to pay for them. Neither party can choose to pull out part way through, at least not without risk of damages for breach. Equally neither can demand more from the other than what he has contracted to provide. While this is the underlying position, such contracts usually contain provisions that modify the position in particular ways related to the nature of building. Several of these bear upon issues which are helped by the preparation of what the JCT contract terms the contract sum analysis, although it does not *require* it for them all. These issues at the post-contract stage are expanded upon in Chapters 9 and 10, but may be listed here to introduce the features germane to setting up the analysis prior to signing the contract:

(a) Changes in design or in other stipulations instructed by the employer. Under most other contracts the former are termed variations, while JCT contracts in particular (including that for design and build) allow also for changes in the latter as obligations and restrictions imposed by the employer. A contract sum analysis is very useful in dealing with the first category by localising how much of the contract sum is open to adjustment because of any change, but not usually of much use in dealing with items in the second category, which do not lend themselves to financial expression as simple entities.
(b) Disturbance effects, the responsibility for which is usually to be laid at the door of the employer. These lead to loss and expense reimbursable to the contractor. While this must often be arrived at by routes other than the contract sum analysis, rather like amounts for changes in obligations and restrictions, it is sometimes possible to use further analysis of the contract sum analysis. This may enable a drop in production, for example, to be assessed or at least be made subject to a test of reasonableness.
(c) Reimbursement of one party (usually the contractor) by the other for changes in market costs, or fluctuations, of particular inputs to production. There are two main ways of assessing these and the 'formula' method cannot be employed without recourse to an analysis. The other method is independent of an analysis, unless it serves as a cross-check on some aspects.
(d) Payments on account of work during progress, or interim payments. It is important that these are adequate for the contractor, who has based his tender on related cash flow projections. It is also important to the employer that he does not overpay by paying too early, as a matter of financial principle, but also because he will be left facing loss if the contractor becomes insolvent and leaves the site. A properly distributed analysis, adjusted during progress by amounts for (a), (b) and (c) above, is most useful when work is valued and paid for at fixed time intervals. It is also useful for calculating in advance amounts to be paid, if this occurs at stages of the physical work, which may be irregular in time.

In addition to these issues which are part of the mechanics of the actual contract, an analysis helps the employer to predict his expenditure flow when set alongside the contractor's programme. This latter document is not required by the conditions, but could be stipulated in the employer's requirements. It is also useful in checking broad progress and in seeing when the contractor is likely to be asking for activity by the employer over various matters laid down in the contract (see Table 7.1). If necessary, the contractor may be asked to provide the employer with an expenditure flow: he should produce the data anyway to predict his own position.

Nature of the analysis

The contract sum analysis is always the analysis of a *lump* sum, which alone is the contract sum, and its nature is governed by the foregoing and compatible purposes which it serves. Without it, later analysis can be contentious, as either party may seek to manipulate figures to his advantage or be thought to be doing so. Even with it, later subsidiary analysis can be subject to the same problem. It is necessary therefore to provide enough analysis, but superfluous (and possibly unwise) to have an excess. This point is expanded under 'Some considerations', below. The broad nature should be laid down in the employer's requirements, but the contractor should always provide it along with his proposals. Even if the employer puts forward any modifications, the contractor should normally remain responsible for having proffered it. The form of analysis may vary:

(a) A number of lump sums. These control 'where the money is': the more there are, the firmer the control. Essentially they may relate to time stages or physical elements (see under 'Structure', below). The two are not mutually exclusive, as elements tend to be sequential in their core content, but one method or the other should give the primary divisions. This appears to be the approach envisaged by the JCT practice note.
(b) A full, priced bill of quantities. This is at first sight very attractive by giving the maximum divisions, showing detailed components of the building structure and so forth, with quantities of each and prices for unit quantities. While it appears to help analysis, it does introduce powerful complexities of its own (see under 'Some considerations', below). Also, because it produces so much detail, it depends upon complete design or upon the suitability of more global bill items which allow for much possibly undecided detail. This can rule it out as the sole method. If it is used, it need not be in accordance with the Standard Method of Measurement and usually is best not so, but more simple.
(c) A number of lump sums, as in (a), with a schedule annexed of unit prices with no quantities, for main components within at least some of these lump sums. Such prices allow changes to be valued, but do not help greatly in dividing up lump sums for interim purposes, as the total quantity within each is initially unknown. It is also difficult for the employer to be sure when entering into the contract that these prices are at the same level as those

making up the lump sums. This can be controlled as in any drawings and specification contract (see Ch. 1), but only by a fairly close reference to the very quantities that are being eschewed. However, if the analysis contains provisional sums for work which is at least partly known, but different in character from that in the rest of the analysis, the existence of contract prices limits agreement for the items concerned to the more objective question of measurement.

(d) Some mixture of the foregoing, with the more suitable method being used for each part according to how advanced is the detailed design. It can also save the labour of a bill of quantities for parts of the works not likely to be changed and simple to divide up for interim payments. It does nothing to resolve the bill complexities mentioned in (b) and hereafter.

(e) With any of the foregoing it is always desirable to have a schedule of daywork rates based upon one or more of the standard definitions agreed with the industry. This is for valuing additional work which is best dealt with on a 'time and materials' basis, because of the difficulty of assessing the level of inputs from first principles.

Structure of the analysis

JCT Practice Note CD/1B acts as a guide for structuring a contract sum analysis under any contract form, whatever is done within the individual sections. The result looks uncannily like the structure of the Standard Method of Measurement as used for building bills of quantities. In that the aim is the same, this deterministic outcome is to be expected. Even if the structure is not adopted wholesale, it is in principle highly desirable. The JCT expect selective use on most projects and further sub-division of some sections on others. Its major sections may be summarised as:

(a) Design work: completed before commencement of construction and during construction.
(b) Preliminaries: site administration, site facilities, temporary work and contract conditions.
(c) Provisional sums: as included in the employer's requirements.
(d) Sections of physical construction.

Some comments are relevant to these elements of structure, so that they are set out suitably in detail as already outlined and as discussed further in Chapter 9.

(a) Design work
The practice note concedes that this element could be included in the value of other elements. It points out that much design work is performed before any construction begins and that it eases early payment for this if it is separately itemised. Later design can also be dealt with ahead of related construction, otherwise the contractor will have to 'front-load' the other elements to provide

early recovery. Against this, the employer may consider that design is of little value on paper; it needs to be translated into physical entities. It should be noted though that an architect usually receives part of his fees ahead of site work.

The other reason for segregating design work is to allow for changes in the employer's requirements, for dealing with provisional sum work and for using the formula method of calculating fluctuations in market costs. In each case the design work costs will not vary in proportion to the others; indeed, for example, if design has been completed and a change leads to a net deduction of work value, the design costs are still likely to show an increase.

The Commissioners of Customs and Excise were consulted by the JCT about the effect on value added tax liability of *showing* design costs separately, as such expenses attract tax when *charged* separately. They stated that liability is not changed, and that tax is payable only when the construction to which the design relates carries tax, so that some new construction is zero-rated. If therefore the works consist of work carrying VAT amounts and work not doing so, the design amounts must be split if shown separately from other amounts. (See also JCT clause 14.)

(b) Preliminaries

This section carries the normal connotation of site costs which are relatively fixed, in that they do not vary directly with the immediate costs of labour, material and plant going into pieces of physical construction. They are thus by way of overheads, best dealt with separately, like design. The term 'preliminaries' does not mean that they are performed first; some removal of temporary work may come last. It is just that specifications and bills of quantities usually make them the first, or preliminary, section!

Preliminaries are a clearly defined section of cost in bills of quantities contracts in which the Standard Method of Measurement applies. That document requires each item to be listed separately, including the titles of all contract clauses. This gives the opportunity for each item to be priced separately, and is not what is envisaged for the present analysis. But often in other contracts the contractor prices some items within the preliminaries and includes the cost of others in his measured prices, out of convenience or necessity. Some do not represent a cost at all. The section taken alone does not necessarily represent costs accurately, as the contractor may choose various distributions within the present analysis. Unless there is some statement of which items are priced as preliminaries, and preferably a breakdown of the total amount, it becomes difficult to use the figure given for contract purposes. A change in restrictions imposed by the employer under clause 12, for example, is easier to value if the original inclusion is known, but is particularly difficult in many instances to include in this isolated way.

(c) Provisional sums

These can occur only in the employer's requirements under the JCT system. In any system, they should be restricted to those agreed to between the parties, both over amount and what they cover. Their isolation in the analysis is tidy, even though they may be expended at different times and places during the work on site.

The work represented by provisional sums should be defined closely enough to avoid uncertainty over their scope. This extends not only to their boundaries and so to the boundaries of contiguous work, but to organisational effects, such as supervision and scaffolding: what are usually termed 'preliminaries'. In quantities contracts it is required that sums for which sufficient detail is given for the work covered are to be classified as being for defined work. This is such detail as the nature and broad quantification of the work and how it relates to the rest of the building and the construction programme. In the absence of sufficient detail, sums are to be described as being for undefined work.

These rules do not apply automatically in design and build contracts, but some similar classification system may be included if provisional sums are very significant in their incidence. It is to be hoped that they will not often be so significant in this type of contract as they may then well upset the design function. Often, it will suffice to state that preliminaries obligations relating to provisional sum work are deemed to be included in the amount or amounts stated under (b) immediately above. Otherwise, there may be discussion over whether the extra costs should be set against the provisional sums, and so increase the final account.

As there are no consultants between the parties in the contract, no general provisional sum to act as a contingency is needed to give such consultants a bit of manoeuvering space within their design. Instead of the employer paying for such frictional matters in the final account, perhaps without really knowing, the contractor must absorb such costs and so should allow for them in the contract sum. It would not suit this type of contract for the designer to have part of the employer's purse available to him. If within the employer's organisation there are those who draw up the employer's requirements and who are responsible to others for the financial outcome, they may consider it prudent to allocate an undefined provisional sum as a contingency. If so, they should control its use by properly instructed changes. In other cases the employer should keep the contingency for his own budgetary use out of the contract sum.

(d) Sections of physical construction

Examples are concrete; roofing and cladding; metalwork; electrical installation; drainage. The first of these, for instance, is then sub-divided into unreinforced concrete, reinforced concrete, reinforcement, formwork, precast units, composite and other special constructions. These are similar to 'work sections' in the Standard Method of Measurement, which are a hybrid of trades and cost elements. Division according to these arrangements may be more suitable in some projects. In any case, the project should be divided primarily into separate buildings, by sub-structure and superstructure, and into external works and cognate units. Each of these should then be divided into work sections or other chosen parts. In very large schemes the allocation of design work and provisional sums in this way might be considered. Preliminaries are usually unsuitable.

Some considerations

Several general issues are implicit in what goes before about the contract sum analysis and some are continued in later chapters. They may be highlighted here, to make them more explicit. They divide into analysis in practice, errors and policy.

Analysis involves the drawings of boundaries between sections of work. This is seldom as simple as 'everything at or below level 56.25', but the use of sections, such as in the JCT list, does mean that well-known and fairly commonly understood divisions are in view. In the ultimate, the contractor includes *everything* in the contract sum and the analysis just shares it out. It is, however, much easier to use the analysis for valuing changes and its other purposes, if both parties are substantially agreed over what is included where, even when some generalised statements are made because of incomplete design.

When the analysis consists of a series of lump sums, there should be enough verbal definition to ensure clarity, but not such as to give the impression that something may be excluded. While it is the employer who should set up the framework of what he wants in his requirements, it is the contractor who must develop it alongside his proposals, as it becomes clear what is contained in his design.

When the analysis consists of a bill of quantities, the statement of content is usually much more detailed and sub-divides the lump sums for the various sections. Here the aim must be to interpret the items given as inclusive of all other items not given, but required to produce what is shown on the drawings, there being no standard method to call to help. The problem of boundaries may still arise, but as part of a much finer grained pattern. The effects of uncertainty over design are taken up in Chapter 9. Although the employer presents the original framework and should ensure that the contractor has complied with it and made the dispositions within the analysis clear, it is the contractor who must be most careful here. If he fails to achieve clarity, the document could well be construed against him, on the *contra proferentum* principle.

The use of lump sums and of priced quantities in the analysis is a distinct principle from that often used in drawings and specification contracts which are not also design and build contracts. Such contracts are intended for projects which are usually fairly small, or at least are relatively simple in content. It is therefore not usually considered necessary to have an analysis, but simply a schedule of rates without quantities for valuing changes or variations. Interim payments are dealt with by post-contract apportionments. An analysis offers information on everything that is *inside* the contract sum and not just on how to deal with changes *at its fringe*. If, however, it offers lump sums for sections and quantities and unit prices within them, it offers a potentially over-rich and even dangerous array of information. The reasons for this are especially grounded in valuing changes and so are most easily discussed in Chapter 9. They should be considered adequately at the pre-contract stage, both by the employer in the form of analysis that he requires, and by the employer and the contractor in looking at the figures that the analysis contains. They relate also to the policy matters mentioned below.

As in any contract, each party should stand by his part of the bargain and so bear, or perhaps gain by, the effects of errors. These are usually made in obvious ways by the contractor, although the employer also may err in assessing the contractor's provisions. The contractor's principal avenues for error are design, quantities and pricing. Design is taken in Chapter 5 in relation to contract clauses. Whether quantities are used in calculating the lump sums in the analysis but not declared, or whether quantities form part of the analysis, the contractor remains responsible for any errors of quantity. If an error is discovered in either direction, no adjustment may be made in the absence of a change instructed under clause 12. Again, the effect of errors on valuing change is covered in Chapter 9. When the contractor provides his analysis, the employer should draw the contractor's attention to any errors of quantity or doubt over interpretation which he may notice. This is best when scheme proposals have met with the employer's satisfaction, but before financial assessment is complete and the contract is formalised. The employer is not obliged to check the quantities and perhaps should not agree them as correct, even if there is time (see Ch. 9).

Any errors in pricing should be treated in the same way, although only gross errors may be apparent to the employer. It is, however, necessary that any arithmetical errors on the face of the analysis be tidied up. Occasionally this may be done most easily by a marginal adjustment of a unit price and extension to keep the same total. Normally, all prices should be left unchanged and the arithmetic carried correctly through to an amended total for either the section or the whole analysis; a percentage adjustment should then be made to bring the total back to give the original amount for the contract sum. If the total of the analysis differs from the contract sum without any amendment having been introduced (and worse things have happened), a similar percentage adjustment should be made. In each case, the percentage will apply in use to each item within the adjusted area. Whatever is done, it is the contract sum that prevails over the analysis to comply with JCT clause 13, or its equivalent.

Policy relates to uses of the analysis other than the proper conventional ones set down under 'Purpose', above. Two instances may be given, one mutual and one solo.

The mutual instance is that the analysis should not be used by either party, even with the agreement of the other, as a means of obtaining 'design to a price' for any part of the works for which there is only outline design at the date of tender. That is, the amount for a section or an apportionment for a part of it should not be used as a budget figure, and so almost a provisional sum, to design what can or should be provided. The employer's requirements and contractor's proposals between them should delineate clearly enough what standards etc the parties intend, and this should be the measure of what is provided. There *is* room for difference here, but so there is in the budget approach which also subverts the package philosophy. If uncertainty forces adoption of the budget approach when the employer's requirements are drawn up, this should be achieved by a genuine provisional sum in the employer's requirements. This will give 'price to a design' when the parties eventually decide on standards etc.

The solo instance is that the analysis may be used by the contractor to distribute the contract sum between sections in a manner to his advantage, unless the employer is able to detect it and insist on a redistribution pre-contractually. Typical examples are loading earlier sections of work to secure finance earlier in the programme, loading sections where it is expected that changes or fluctuations will produce increases at a favourable rate, and lightening them where the opposite is expected. If quantities are introduced within sections, there is a further relationship between them and their unit prices, as considered in Chapter 9, and the contractor may seek to exploit this. Such action can be risky if forecasts are wrong, but both parties are bound by the distribution once the contract is in being.

Financial basis: the JCT contract

As already discussed, this form is an 'entire' contract, with the works and the lump sum price constituting two balancing elements of consideration in the legal sense. These clauses spell this out, while allowing for balancing adjustments of the two elements. Anticipated revised section headed clause numbers are given in brackets (see also the Index of JCT clauses).

JCT clause 13(4.5) – Contract sum

As this clause is so basic, it is a constant disappointment that it always occurs so late in every issue of JCT conditions, and particularly after clause 3. It establishes the contract sum as fixed, unless 'the express provisions' in the contract allow otherwise. Various provisions are scattered throughout the clauses, some authorising 'adjustment' of the contract sum and some allowing deductions or similar from the otherwise adjusted sum. Sometimes rules for arriving at the financial results are given but, in the nature of some of the cases, this is not always practicable. The various varieties are dealt with under the clauses concerned, but comprehensive lists are given in Table 6.2.

With all these listed, there remains the possibility of the two parties agreeing to some further adjustment, as parties to a contract may always do. In respect of the laid down adjustments, arbitration is usually provided over failure to agree. This may be more necessary in the absence of an architect and a quantity surveyor to mediate between the parties, although their absence may simplify negotiations, if only because of ignorance! The long-stop is still court action.

What is absent from the area of potential adjustments is any rectification of arithmetical errors in computing the contract sum, whether these be quantitative,

Table 6.2 Financial adjustments provided in the JCT with design form

Express provisions referred to by clause 3

Clause	*Subject*
6.2	Statutory fees or charges
8.3	Cost of inspection and testing
9.2	Royalties arising out of instructions
12.6	Change and provisional sum expenditure*
16.2	Making good defects
16.3	Making good defects
22B.2	Insurance premiums on employer's default†
22C.3	Insurance premiums on employer's default†
22D.4	Insurance premiums regarding liquidated damages
26.3	Loss and expense due to disturbance
34.3.1	Loss and expense due to antiquities
36.3.2	Fluctuations
36.4.4.1	Fluctuations
37.4.2	Fluctuations
37.5.4.1	Fluctuations
38.1.1.1	Fluctuations

* *See also list of 'deemed' changes in Table 7.2.*
† *Not applicable in local authority contracts.*

Clause 30.5.3 also falls under clause 3. It gives a list of final adjustments and again covers all the above.

Deductions or recoveries by employer

Clause	*Subject*
4.1.2	Work by others on contractor's non-compliance
21.1.3	Insurance premiums on contractor's default
21.2.3	Insurance premiums on contractor's default
22A.2	Insurance premiums on contractor's default
24.2.1	Liquidated damages (repayable under clause 24.2.2)

Miscellaneous provisions

Clause	*Subject*
27.6.6	Balance after determination by employer
28.4.3	Balance after determination by contractor
28A.5	Balance after determination by employer or contractor

Payments outside main framework

Clause 31 Statutory tax deduction scheme
Supplementary provisions VAT agreement
Clause 14.3 refers to payment for loss of credit for VAT, without defining how this is to be made.

financial estimating or simple arithmetic. In the case of the JCT with quantities forms, an adjustment may arise because of errors in the preparation of the bills of quantities, although not because of errors in pricing them. This is because the bills are proffered by one party for use by the other (a practice discouraged under the present contract in comments in this chapter and in Chapter 9). Because of this limited scope for adjustment, these other forms need expressly to exclude other adjustments. In the absence of the former, the latter is not needed either in the present forms. Discrepancies within contract documents and divergences between them are a distinct issue and are covered by clause 2.

JCT contract sum analysis

The principles of a document of this type are discussed earlier in this chapter. Within this contract its total mention is:

(a) Second recital, as being submitted by the contractor with his proposals.
(b) Article 4, as being signed by the parties and identified in appendix 3.
(c) Clause 12.5, as the starting point for valuing changes.
(d) Clause 38.2, as being amplified by the contractor for implementing formula fluctuations.

There are notes on the contract sum analysis in JCT Practice Note CD/1B, where it is indicated that the analysis is intended to facilitate work by both parties over amounts of interim payments when the periodic alternative under clause 30.2B is used. This is not a contract stipulation, so that neither party can insist on using it in this way. The analysis is, however, prima facie obvious data and normally will be used. It is also reasonable for the first alternative to tie up with it so far as is suitable.

The practice note gives a guide list of work sections for the analysis, adjustable according to the scale or complexity of the project. Comment on the technicalities of such a list, and on the question of errors, is made earlier in this chapter.

JCT clause 3(4.3) – Contract sum – additions or deductions – adjustment – interim payments

The purpose of this clause is to permit progressive account to be taken of adjustments to the contract sum. These will be either additions or deductions, but those terms are also used in their own right in the conditions, and so here. All the authorising clauses are listed in Table 6.2, but the present clause fails to refer also to other financial adjustments listed here. These sometimes have their own rules, but are not 'adjustments [etc.] of the Contract Sum' as referred to here.

Amounts may be taken into account after 'whole or partial ascertainment'. This allows for approximate calculation of the whole or even part, as well as precise and final calculation of amounts for inclusion in interim payments. In the event of a disagreement, an early arbitration is possible under article 5.2.1.

Chapter 7

Construction, control and quality

General administration of the works
Persons engaged upon the site or the works

This chapter is concerned with the major phase when building work is proceeding on site. It stops short of completion and what lies beyond; these are covered in Chapter 10. Discussion is given of those clauses in Section 1 of the anticipated revised section headed JCT contract not discussed in Chapters 5 and 6 and of the clauses in Section 3 of that form (revised numbers are in brackets and see also comparison table in the Index of JCT clauses). Most of the details are brought in by the contract clauses discussed, but some key activities of the two parties may be listed as pointers, including some discussed in later chapters.

(a) *The contractor must*:
  (i) complete the design
  (ii) carry out the construction to the contract standards for materials and workmanship and in conformity with statutory requirements
  (iii) provide a person in charge of the works
  (iv) complete on time (although perhaps not unduly early) subject to arrangements for extending the time for given reasons (see Ch. 9)
  (v) deal with contingent matters, such as antiquities discovered (see Ch. 9)
(b) *The contractor may*:
  (i) organise the execution of work as he wishes, subject to any constraints in the contract
  (ii) assign or sub-contract this execution, subject to the employer's approval
  (iii) object to unreasonable changes in the design sought by the employer
  (iv) seek payment for disturbance, as well as changes and other provisions (see Chs. 8 and 9)
(c) *The employer must*:
  (i) give possession of the site on time
  (ii) act over discrepancies and divergences affecting what is to be produced

(iii) give occasional consents etc during progress, but should not do so in a way that relieves the contractor of his responsibilities
(iv) grant extension of the contract period when appropriate (see Ch. 9)

(d) *The employer may*:
(i) appoint an agent and others to represent his interests, who may have access to the works etc
(ii) assign his rights, subject to the contractor's approval
(iii) give instructions on specified matters only, these including changes in design and particular obligations and restrictions and postponement of work (see Chs 8 and 9)
(iv) have other persons performing further work on site at the same time as the works, provided the contractor has agreed pre- or post-contractually

General administration of the works

Several JCT clauses etc may be grouped under this rather amorphous heading. They are those which deal essentially with straightforward site operations on a day-to-day basis.

JCT article 3 – Employer's agent

The reference here is to a person whom the employer nominates to act on his behalf during the contract period. Even if he has acted earlier, this person must be nominated separately for his post-contract activities. There is only one agent at any one time as the employer's immediate representative, although another may be introduced 'in his place' by the employer. There is no reason why the agent may not designate his own deputies for particular activities, and for specialist functions he usually should. There is a limit to interchangeability, for instance, between design consultants, quantity surveyor and clerk of works.

An unqualified nomination will invest the agent with all the functions of the employer under the conditions, by virtue of the comprehensive list 'for the receiving or issuing . . . or for otherwise acting for the Employer'. These activities will therefore include, for example, receiving applications for payment, consenting to assignment or sub-letting and issuing a notice of determination, variously under clauses 30, 18 and 27. Without qualification, the list even includes paying the contractor! Several of these examples are well beyond what the architect may do under other JCT forms, where he is in an equivalent position to the agent.

The employer should always 'otherwise specify' when nominating to restrict the agent's responsibilities, although the precise restrictions may vary according to the nature of the works and the expertise of the agent. When particular roles are not those of the agent personally, or of his organisation, the employer may be advised to instruct the agent on those which he requires to be delegated to named persons. All of this may be easier if the agent is a member of the employer's own organisation, although here the necessity for letting the contractor know the extent of delegation becomes even stronger.

A list of the functions allocated to the employer by the conditions is given in Table 7.1. In the absence of an agent the employer will have to perform those that are mandatory, and also such of the rest as he does not wish to go by default. Only in clauses 5, 4 and 11 is there mention of the agent, and here the employer strictly must have nominated *himself* as agent to be able to look at documents on site or to have access to the works and other places! In all cases the employer should weigh the consequences of having no agent and, if he has one, of how duties are to be divided. The agent in turn needs to realise that he is an *agent* and that he may not be acting in the same way as under other forms of contract, nor have the same authority.

For simplicity the comments on clauses in all chapters refer to the employer alone except in a few instances, but the term 'employer' should be read as including the agent where this is reasonable.

JCT clause 1(1.1. and 1.12) – Interpretation, definitions etc

All of the terms, such as 'Contractor' and 'Development Control Requirements', are listed here in the usual JCT way and either defined or referenced to another clause which does define them. Their significance, where not obvious, is dealt with under the appropriate clauses.

JCT clause 5(1.6 and 1.5) – Custody and supply of documents

This clause resembles the other JCT versions, but gives a two-way supply of documents under clauses 5.1 to 5.4. Thus, while the contract employer's requirements, contractor's proposals and contract sum analysis are kept by the employer and he supplies copies to the contractor, as in those versions, it is by contrast the contractor who is to supply the employer with further drawings etc during progress for the simple reason that the contractor produces them. It is a little odd, though, that the employer has to supply to the contractor documents which may include a significant number of drawings which the contractor has prepared in the first place and he alone is entitled to amend. In keeping with the contract philosophy, there is no mention of the employer approving the further documents available for reference by the other. The contractor's set is to be on site as is usual practice and includes the contract sum analysis. The availability of priced information on site, as required here, is not so usual. Strictly, the site set is available to the employer's agent (see article 3), rather than the employer himself, according to the wording.

Clause 5.5 is peculiar to this contract and requires the contractor to supply even more drawings and information to the employer, 'Before the commencement of the Defects Liability Period', that is, not later than practical completion on site, so that he has to be pretty quick in the closing stages. A failure does not lead to

Table 7.1 Main rights and responsibilities of employer under JCT with design form

Article	*Subject*
2	Consideration for works
3	Appointment of agent
5	Arbitration (referring to clause 39)

Clause	*Subject*
2.4.2	Discrepancies in contractor's proposals
4	Instructions (see Table 7.2)
5.2	Provision of documents
6.1.2	Divergences from statutory requirements
7	Definition of site boundary
11	Access for inspection
16	Practical completion notice and defects liability period
17	Partial possession of works
18.2	Consent to sub-letting of construction and design
21	Insurance of liabilities
22	Insurance of works
22C	Determination after damage
23.1	Give possession of site for works
24	Non-completion and liquidated damages
25.3	Extension of time for completion
27	Determination by employer
28	Determination by contractor
28A	Determination by employer or contractor
29	Work not part of contract
30.1	Interim payments
30.5.5	Check contractor's account and statement
30.5.6	Prepare account and statement on contractor's default
30.6	Final payment
30.8	Arbitration or proceedings
31	Statutory tax deduction scheme
34	Antiquities

Several of these items are covered in several related parts of the clauses listed, and these should be consulted. Insurances and payments in particular have extensive ramifications. In other JCT contract forms, many of these matters are administered by the architect or quantity surveyor.

liquidated damages (see clause 24), but might justify an extension of the defects liability period if the employer could not check the works properly in the normal period. What he has to supply must be stated in the contract documents, otherwise the clause is ineffective. Documents 'describing the Works as built' act to tidy up what has been provided progressively during construction. Those 'concerning the maintenance and operation of the Works' may be manuals or recommendations and their extent is really dependent upon the nature of the works. Both sets may include 'any installations', meaning such categories as plumbing, heating, ventilation and electrical systems. Even in the simplest of buildings, the employer should require and check information about where concealed runs etc are to be found and how to control the installations and keep them efficient.

While there is not the rather outdated requirement for the return of documents to the other party, clause 5.6 does oblige the parties to maintain confidentiality and also not to use any documents for extraneous purposes. This affects such matters as the employer's use of the works and perhaps commercial data, and the contractor's design and pricing. The particular aspect of copyright falls under clause 9, considered in Chapter 5, and would cover documents mentioned in clause 5.5 and not strictly covered by clause 5.6.

The exception in the clause is that the employer may use documents supplied to him for 'the maintenance, use [and] repair . . . of the Works' (which may appear quite obvious), but also for 'the . . . advertisement, letting or sale' of them. This is reasonable in facilitating these activities and in saving production of information by amending pointlessly what is already in existence. There would not appear to be any circumstance in which the contract sum analysis should be passed over. 'Advertisement' is a public matter and care is needed here not to put about confidential aspects of design, such as specialised processes. The clause does not expressly allow the employer to provide documents to any tenant or purchaser for his own use and clause 18.1.2, if in use, is not explicit on the point, although an assignment might prove difficult to implement without such provision. To avoid doubt, the employer should include any known matters in the employer's requirements.

JCT clause 4(3.5) – Employer's instructions

This very important clause is the equivalent of that entitled 'Architect's instructions' in other JCT contracts. It deals with the procedures over instructions in terms almost identical to those in similar clauses, virtually by substituting 'Employer' for 'Architect' throughout. By article 3 the employer's agent, when one is appointed, is to be read into this clause (as into others) in place of the employer. The employer should, and indeed must, act only through his agent unless he limits his terms of reference as noted under article 3, and if this clause is read closely in this light it illustrates well the role of the agent in relation to the contractor. However, in relation to the employer the agent may only pass on instructions to the contractor and not originate them, unless his appointment with the employer permits otherwise. The scope of instructions under this contract is rather different from that under others, as the list in Table 7.2 indicates, and they have a rather different status also by virtue of being issued by a party to the contract and not by the architect who is acting impartially between the parties.

While the mechanics of the clause are like those in the JCT standard form, some

Table 7.2 Employer's instructions etc to contractor under JCT with design form

Formal instructions

Clause	*Subject*
2.3.1	Divergence over employer's requirements and boundaries
8.3	Opening up and testing
8.4	Removal of defective work etc
12.2	Changes
12.3	Expenditure of provisional sums
16.3	Defects during defects liability period
22D.1	Insurance over liquidated damages
23.2	Postponement of work
34.2	Antiquities

'Deemed' changes

Clause	*Subject*
6.3	Statutory obligations
22B.3.5	Restoration after damage and removal of debris
22C.4.4.2	Restoration after damage and removal of debris

Other matters

Clause	*Subject*
7	Definition of site boundaries
16.2	Schedule of defects
27.2.1	Notice of default
27.2.2	Notice of determination
27.6.3	Removal of plant etc
28A.1.1	Notice of determination

points may be emphasised. Clause 4.1.1 requires the contractor to comply with all employer's instructions, but limits the scope of instructions to what is 'expressly empowered by the Conditions'. This means that the employer cannot step outside the list without risking a challenge under clause 4.2 to substantiate his right to do so. If the contractor complies with the instruction without reservation after such a challenge, but only so, he and the employer are bound by it without right to arbitration (the effect of 'for all the purposes of this Contract'). He may, however, seek arbitration and so reserve his position. In the cases given in clause 39.2, arbitration need not be delayed until practical completion. This may be crucial over design changes, if the employer seeks to force a solution on the contractor under clause 12.1.1. In respect of changes in obligations or restrictions under clause 12.1.2, clause 4.1.1 gives the contractor a right of 'reasonable objection' to an instruction which would seriously disrupt his working operations, rather than his design. In both these cases, arbitration may proceed at once.

The contractor may neglect or even refuse to comply with an instruction, without seeking to use clause 4.2 to clarify matters. If he does, the employer may give him notice under clause 4.1.2 to comply and thereafter, if the contractor still does not, may introduce others to carry out the parcel of work concerned and recover the cost from the contractor by one of the methods specified. This is straightforward over non-removal of defective work or even performance of fresh work on site, which is the normal point of the provision. It becomes more contentious if the work which the contractor has failed to perform contains or consists of design and especially if this design overlaps with other design, whether carried out so far or not. There is not a right for the employer to seek immediate arbitration nor, unless *all* outstanding design work stops, to give notice of determination. It is difficult to envisage how such an impasse might develop, but clearly the contractor is in the wrong if it does, even if only by neglecting his proper remedy under this clause. The employer should be entitled to proceed with caution, leaving the contractor to absorb the results into his design and within the contract sum, without adjustment.

The employer may otherwise have sought to enlarge the scope of his instructions by extra stipulations in the employer's requirements. These are not under any 'provision of the Conditions' as the present clause requires and so offend against clause 2.2, making them ineffective. In any of these cases the contractor may be prepared to comply, but he is entitled to do so only upon such special terms of payment as he may stipulate and have agreed before he acts.

Clause 4.3 contains the usual formula that instructions are to be in writing initially or otherwise are to be confirmed by one of the parties: by either of them before the contractor acts, or by the employer thereafter. In no case need the contractor comply until the instruction is either given or confirmed in writing, so that none of the procedures over his failure to comply come alive until the written form is present.

Alternatively, the contractor may comply with an oral 'instruction' which does not receive confirmation. This may be because the employer disputes that it was given, and matters will turn on the plausibility of evidence. Under other versions of the contract, no financial adjustment up or down is due unless instructions are given or confirmed in writing and the quantity surveyor or architect will *exclude* any allowance from the final account. This should alert the contractor to his need to seek confirmation, especially if the adjustment is upwards! In the present contract, the need for written authority is technically the same, but the contractor can provoke matters by including an adjustment in the final account. The employer may then, if in agreement, choose to confirm the instruction or possibly to acknowledge it by simply accepting the item in the account. The former and more formal course is desirable if the possibility of evidence at a later stage exists and the account is imprecise over details. Clearly the employer should also look for instructions not taken in the reckoning, perhaps because they are unconfirmed and result in a deduction. These aspects relate to the final account considered in Chapter 10.

JCT clause 7(3.8) – Site boundaries

The employer is responsible for defining the boundary, whether by pegs, fencing or drawings of sufficient accuracy. This leaves responsibility for setting out with the contractor. If the definition diverges from that in the employer's requirements, clause 2.3.1 allows a change to apply under clause 12, with the usual possible remedies following.

JCT clauses 8(1.2 and 3.11 to 3.13) – Materials, goods and workmanship to conform to description – testing and inspection

The essence of this clause is to provide for the employer receiving what he has contracted to receive and checking that he is getting it. This is done in six parts similar to those in the JCT standard form, there being no equivalent here allowing the employer to have anyone excluded from working on site.

Some similarities may be summarised. All materials and work by clauses 8.1.1 and 8.1.2 are to be as set out in the contract documents and the employer may request vouchers under clause 8.2 to prove this. In addition to what the JCT standard form provides, he is entitled to samples of workmanship and materials under clause 8.6, to the extent that these have been specifically referred to in the contract documents. How much all of this means will depend on how far the documents give explicit standards and how far they are limited to performance specification. The latter may not be susceptible to testing by taking samples piecemeal, but only as part of the finished operating works.

Further, as parts of his general right of checking, the employer may instruct under clause 8.3 that any covered work should be exposed or that any work or materials be tested at any stage. To make this practicable, clause 11 gives rights of access to work etc. There is in principle no limit on how late in the programme the employer may have work uncovered and how much disturbance he may cause, nor on the frequency of tests. The employer's enthusiasm will be tempered, however, by the knowledge that liability to meet costs incurred will lie with him, unless there is non-compliance with the contract. Where tests are of a routine, predictable nature by way of quality control, such as concrete testing, they may be spelt out in the contract documents and the contractor must include for them in the contract sum. Here again requirements should be reasonable, otherwise the contractor must make an unreasonable allowance when tendering. Once default is established, the employer may instruct removal or rectification under clause 8.4 of what is defective, so that the contractor is then bound to act without charge, so as to meet his obligation of completing the works. If he does not, the employer has the option in a serious case of employing others to do the work under clause 4.1.2, or in a drastic case of determining the contractor's employment under clause 27.2.1.3.

Clauses 8.1.3 and 8.5 deal with the question of the contractor carrying out the works 'in a proper and workmanlike manner' or failing to do so. The wording follows the standard form and so should the effect. The employer may need to tread a little more carefully before invoking clause 8.5 if the contractor is using specialised construction of his own choice, although it was negligence in such a case which led to the somewhat unexpected introduction of the present clause.

There are several points of difference from the JCT standard form. Due to the nature of the contract, there is expansion of clauses 8.1.1 and 8.1.2. This allows firstly for the various kinds or standards being as described in the employer's requirements or otherwise in the contractor's proposals (corresponding to contract bills or specification), or possibly as in the post-contract specifications mentioned in clause 5.3. In all cases they are to be provided 'so far as procurable'. The latter phrase in all JCT forms leaves the same doubt as to whether it means 'has ceased to be procurable since the date of tender', rather than that there was an initial contract to perform the impossible. If the item was in the contractor's proposals, it allows the possibility in a dubious case of the contractor having misled the employer, even if accidentally, in this perhaps more fiduciary contract. The clause makes no distinction over how the difficulty arises, including in post-contract specifying, and thus substitution of an equivalent item becomes inevitable. The proviso in the clause about making a substitution is quite general though, and is not contingent upon this 'procurable' question. Thus the contractor may also wish to introduced an alternative to what he or the employer originally specified for some other reason, perhaps price or speedier delivery, while still providing a similar quality of finished building.

Whether the substitution is forced or voluntary on the contractor's part, he has to obtain the employer's consent to what he proposes and this consent is not to be 'unreasonably delayed or withheld'. The employer cannot therefore insist on the unobtainable, but further he may be unreasonable in not consenting to a substitution of another 'kind' which is still of the same 'standard', in that it meets his requirements when expressed as a performance specification. If he provided the contract specification he may well not have used that mode of specifying, but it could be applied as a test of reasonableness, unless his requirements allowed only one way of being met and that way is still possible. If the contractor provided the specification in a performance mode, and this was accepted, the employer's consent is not needed to anything which meets that specification. When the contractor has once provided any firm specification under clause 5.3, however, he is obliged by this present clause to seek consent before changing it.

Nothing is said here about any adjustment of the contract sum and, provided that the employer's requirements are still met, there should be none whether the item as performed costs the contractor more or less. This is in the spirit of this contract just as much as a difference in costs due to sub-letting in building contracts in general. There is not a change in the employer's requirements under clause 12, leading to difference in payment. The present clause relates to changes in the contractor's proposals without antecedent changes in the employer's requirements. In the event of impossibility of procurement in both 'kind' and 'standard', the employer will be forced to concede and instruct a change in his requirements (even if the detailed specification is in the contractor's proposals), so that a financial adjustment will follow. It may be that in some borderline cases over what is possible for the contractor or reasonable on the part of the employer, the parties will agree to a substitute with some measure of financial adjustment, but in this the contract is silent.

In a second expansion, clause 8.1.1 states that the employer's consent is not to 'relieve the contractor of his other obligations'. As discussed in Chapter 5, the employer consents to the contractor's proposals and to his later developments or modifications of them and should be careful not to approve them in such a way as to relieve the contractor over such matters as design, performance or quality. The comments under clause 30.8.1 about the employer's 'reasonable satisfaction' should also be consulted.

Clause 8.4 differs from the standard form by adding 'rectification' to 'removal', as would be reasonable in that form. It contains three alternative actions (which may be taken in combination) flowing from the discovery of what is not in accordance with the contract: to instruct removal or rectification, to instruct a change or to instruct further opening up or testing regulated by the code of practice appended to the conditions. These are as in the JCT standard form, but the possibility of allowing work to remain subject to a deduction from the contract sum is not given, the reason for the omission being unclear. In view of this omission, it is also unclear what sort of change might be instructed if all that may be required of the contractor is compliance with the contract unless, for example, some redesign to regain time is in view. Once more, the whole procedure is likely to be affected by the extent of performance specification used in the contract documents.

JCT clause 15(1.9) – Unfixed materials and goods

This clause repeats the first part of the JCT standard form provision, which says three things:

(a) Materials on site are not to be removed without the employer's consent.
(b) Materials included in an interim payment, that is, when the employer actually pays the contractor, become the property of the employer.
(c) Responsibilities over loss and damage are not altered by payment.

The second part of that provision relates to materials not on site. It occurs here as part of the optional provisions in appendix 2, available when the alternative of periodic payments applies under clause 30.2B, where it is discussed. The position does not apply when stage payments under clause 30.2A are used, although it could be introduced by agreement. The reference is to property passing when items are 'included in any Interim Payment'. When periodic payments apply, this is usually clear from the detailed computations. When stage payments apply, the value included for materials is some average expectation allowed when the overall stage values were calculated. Unless therefore the actual materials are recorded when a stage is reached and are expressly agreed to be included, no effective transfer will occur.

JCT clause 10(3.4) – Person-in-charge

This is the standard JCT requirement for the contractor to have someone regularly on site to run construction of the project and to receive instructions, but not other forms of contract communication such as notices and consents. The employer should therefore take care over who receives his various missives, if they are to be effective contractually. The contractor may wish to designate his person-in-charge to receive some further documents, such as consents, in the interests of direct communication. While some instructions will deal with matters like changes in design which are not to be settled by the person-in-charge, the contractor is advised not to try to separate them out, but to leave them all to come through the site rather than some to site and some to head office. Certainly he should not allow them to go direct to any consultant of his.

JCT clause 11(3.16) – Access for employer's agent etc to the works

While the contractor has exclusive possession of the site, subject to clause 29 about allowing other persons to work there, it is necessary to secure access for inspection and so forth by the employer. This clause does this and extends the access to the contractor's workshops etc without qualification and also to those of sub-contractors 'so far as possible'. If Sub-Contract DOM/2 is used, this gives a right of access by its clause 25. The sub-contract is optional and the clause can be deleted. Whether the right of access is 'possible', or even useful, depends perhaps upon the sub-contractor, commercial secrets or whether goods can be identified in production line work.

Who secures access is a little confused. To the works and the contractor's workshops, it is the employer's agent and others 'authorised' by him or by the employer. To sub-contractors' workshops, it is the employer and his 'representatives'. If there is any awkwardness, presumably those who can will endorse those who otherwise cannot.

'Places . . . where work is being prepared' in this contract includes design offices and laboratories, as it does under nominated sub-contracts to other contract forms when sub-contractors are designing.

JCT clause 34(3.18) – Antiquities

These are unpredictable contingencies. The clause also covers 'other objects of interest and value', usually to be found during excavations. These become the property of the employer and the contractor is to minimise disturbance, as the position and condition may be crucial. The employer has power to issue instructions and have a third party come along to deal with objects. This does not strictly constitute a change to be valued under clause 12, but the contractor is to be reimbursed his loss and expense in a similar way to that in clause 26. Extension of time may arise under clause 25.

Persons engaged upon the site or the works

Not only is the organisation of the works generally in the contractor's hands, so too is the question of who actually does work and how he relates to them. These clauses introduce some restrictions and also the possibility of others doing work on site, not as part of the works and over which he has no control. A further possibility arises with statutory undertakers recognised in clause 6, as discussed in Chapter 5. The clauses are not peculiar in principle to this JCT contract, but warrant some comment.

JCT clause 18(3.1 and 3.2) – Assignment and sub-contracts

Assignment of the contract itself is dealt with in clause 18.1 by the usual simple prohibition on either party acting without the consent of the other. Also corresponding to the JCT standard form is clause 18.2 about assignment of the employer's interests after practical completion, and so after most contractual obligations have usually been fulfilled, to those to whom he may sell or lease. This effectively removes the employer from being a middleman, while not allowing the assignee to dispute agreements reached before the date of assignment. Reasonably, the contractor does not have a right of objection to any assignment in this latter situation. This area is complicated by the effects of any collateral warranties required of the contractor or sub-contractors, as considered in Chapter 10.

Clause 18.2 is about sub-letting, that is the introduction of what are usually called domestic sub-contractors, whose contractual relationship is solely with the contractor. It distinguishes 'the Works' from 'the design ... for the Works', whereas some clauses do not. Over both elements it makes the normal provision requiring the employer's consent to sub-letting 'not ... unreasonably delayed or withheld'. In the case of the works only, there is included the provision about automatic determination if the contractor's employment is determined (see Ch. 9), to avoid any carry over of the employer's liability. This is omitted in the case of design, presumably in the hope of leading in a negative sort of way to some agreement over continuity in this critical area. This divergence of treatment suggests an attempt both to retain and to assimilate the cake; with at least one activity legally flawed.

Sub-Contract DOM/2 is available to the contractor for use in any case of sub-letting, but is quite optional. It incorporates and amends the terms of Sub-Contract DOM/1, which read alone is used in relation to other JCT forms. Both are worded as site work sub-contracts, although Sub-Contract DOM/2 introduces design. The result is not really suitable for either a design only sub-contract or a fabricate and deliver only sub-contract.

Sub-Contract DOM/1 provides for automatic determination of the sub-contractor's employment with that of the contractor, and is not modified when brought in here. This nullifies the distinction introduced into the present clause. It also allows the sub-contractor to sub-let with the consent of the contractor alone, so nullifying another aspect of the clause.

Sub-letting always holds a risk element for the contractor, in that no adjustment of the contract sum occurs if the contractor cannot sub-let for any reason to a person he had expected when tendering. This risk is potentially greater when design is involved, either as the whole or as part of the content of the sub-letting. Not only may the contractor wish to sub-let such design post-contractually, which is what the clause is dealing with, but he may already have sub-let the earlier stages pre-contractually. He is most likely to want continuity here. If he is performing most of the total design of the works in-house, it may present relatively little difficulty if he has to switch to another person over the design of just a self-contained part of the works. If he is having all design performed by an external consultant, he may be heavily dependent upon retaining him to carry the design through in concept and detail, affecting its cost in time and construction. If the external designer is also the prospective constructor or installer, there is a sub-letting to consider under both parts of the clause. But often it will not be possible to take the existing design concept and transfer it for another person to develop and then produce physically, owing to questions of proprietary components and so on. A switch may therefore change the design far more radically and may also lead to a quite different price level for site work as well. If an intended or approved designer withdraws, the issue is raised again, perhaps at an unexpected and critical juncture.

In view of these elements, the reasonableness of the employer withholding consent is under more strain over design than over construction (as also happens with the situation of determination). Yet it is here that he too may have more to be concerned about, as the design is basic for all else. Both parties are therefore well advised to see that the major sub-lettings are cleared before the contract is formalised to avoid later contention (see Ch. 6 over design consultants). This should not prevent the contractor from coming forward with alternatives post-contractually if he so wishes, but he does so then after a good opportunity to reduce his uncertainties. In all cases, the clause will apply by leaving the contractor's design obligation undiminished. He may make what arrangements he will with other designers over liabilities, the employer can and will look to him alone for any redress in contract, although action against these others in tort may be possible (see 'Defects after settlement' in Ch. 10).

There is nothing in the clause about the employer stipulating who shall perform any part of the design as a sub-contractor. With the main design, this would obviously destroy the design and build concept. With self-contained and essentially free-standing elements, in principle it is possible within the broad concept. In this contract, it could be achieved by naming one or more persons in the employer's requirements and leaving the rest to the contractor, or by including a provisional sum for the element concerned and stipulating that the employer will name someone to perform the design work involved. The contractor's designed portion supplement (see Ch. 12) is used in conjunction with JCT standard forms which specifically provide the option of a list of not less than three domestic sub-contractors from whom the contractor may choose. This therefore admits the system into the design work itself. While many things are possible, not all are expedient and the employer should consider the dangers of splitting design responsibility as discussed in Chapter 5.

JCT – clause 29(3.10) – Execution of work not forming part of contract

This clause is a replica of other JCT versions, subject to substituting the employer's requirements for other documents. Both its title and its contents need the addition of 'on the site' to make explicit its inferred purpose.

The clause needs even more care over its implementation in this contract. Any work introduced by it should be quite distinct from what the contractor does, such as late embellishment or separate work in the grounds. Work which is integral with other aspects, such as a structural frame or a heating installation having design and performance repercussions, should never be treated this way (see Ch. 5). It allows either the employer or others on his behalf to perform work contemporaneously with the works themselves and so to have access.

The contractor simply has to 'permit the execution of such work' and any person is declared 'not to be a sub-contractor' by clause 29.3, so that the employer is responsible for all dealings, including co-ordination and payment. The clause relates only to 'work not forming part of this Contract', so that it is not possible to take any work from the contractor and give it to another. This would infringe the entirety and design and build principles and again lead to divided responsibility.

Two ways of introducing work are given. In clause 29.1, there is the case of the employer's requirements giving enough information 'to enable the Contractor to carry out and complete the Works', that is, enough to enable him to plan his operations around the other work and allow for this when tendering. There is no requirement for him automatically to provide any facilities such as access to scaffolding or attendant labour etc. Anything at all which he is to provide must be given, otherwise he is simply being warned of what is to happen and how it will affect the works. If the design is also affected by say superimposed loads, the requisite information must be provided. Given enough detail, the contractor must allow this other work to proceed.

In clause 29.2, there is the case of the employer's requirements not giving the required information. Thus inadequate information may be given, or there may be no mention at all of the extraneous work. In either case the employer may arrange for the work only with the contractor's consent, not unreasonably withheld. Thus the employer cannot secure his position in advance by a nominal inclusion of the work without detail. The contractor may still be able to decline it happening. Concerning matters of design, it is possible for 'unreasonably' to be interpreted more narrowly under this contract in the contractor's favour.

These considerations point up the virtues of the employer being precise and fulsome. So too do the remedies otherwise available to the contractor by way of extension of time, loss and expense payment or determination if disturbance results. Even if he allows a person on site, he may well wish to establish his remedies before doing so. The clause does not govern relationships over supply of materials (where similar remedies may be invoked), and the employer's requirements must give all information about supply and reception, as well as specification and quantity control. Here too the dangers of divided responsibility lurk in the shadows.

Chapter 8

Progress and disturbance

Programme of the works

Loss and expense

Injury, indemnity and insurance

Determination of the contractor's employment

This chapter deals with the provisions of the relevant JCT clauses, which in most respects are direct reproductions of those in other JCT contracts. They constitute sections 2, 6 and 7 of the anticipated revised section headed JCT contract (revised numbers are in brackets and see also comparison table in the Index of JCT clauses). Incidental mention of design is made in a number of places, while in others the contractor's design responsibility affects interpretation. In view of these points and the generally complex nature and length of the clauses, a rounded and fairly detailed treatment is given.

Even so, it is significant that no special provisions are given about progress as such. This is indicative of the difficulty of putting such matters into words, but also of the way in which the vaunted speed of design and build is not dependent upon explicit provisions but, when present, derives from the 'one organisation, one process' concept. Introductory remarks are not therefore needed for this chapter, although a review of sections of Chapter 2 will bring forward key background matters.

Programme of the works

While the contractor may organise the works however suits him best, subject to any constraints laid down in the employer's requirements, he has to work between the dates for commencement and completion. The group of clauses in this section deal with him doing or failing to do this, while events at and following completion are considered in Chapter 10.

JCT clause 23(2.1, 2.2 and 3.14) – Date of possession, completion and postponement

The date of possession mentioned in clause 23.1.1 is inserted in the contract appendix. If the employer is late in affording possession, he stands to be in breach and the contractor has the usual remedies open to him: in a serious case, of repudiation and suing. More temperately, he may accept the situation with any adjustment of the contract sum which is needed and which he negotiates in consideration of carrying on. Clause 23.1.2 is optional, but if in use allows the employer to defer possession of the site by the contractor for up to six weeks. Desirably this should be done by giving as much notice as possible. In default of adequate notice, the contractor's entitlement to extension of time or loss and expense payment is likely to increase. In the normal course of events, the contractor is required to 'begin the construction' upon the date of possession. This presupposes that he has taken his detailed design to the point where site work is possible, for which he must allow accordingly in his programme. His obligation, logically or otherwise, is to start on site on that date and proceed 'regularly and diligently'. The only relaxations he has over starting or proceeding are as provided in clause 25.

One of those relaxations is over postponement of design or construction by the employer, permitted in clause 23.2. This may affect all or part of the works and does not have to be 'reasonable'. The contractor must accept it without arbitration over its instruction, subject to any extension of time or loss and expense payment that may be justified. Only in an extreme case may he determine his employment (see clause 28.1.2.4). If, however, he suspends all work or fails to maintain proper progress he may risk a determination against him (see clauses 27.1.1 and 27.1.2).

The completion date is defined in clause 1 as the date of completion (surprise) in the appendix, that is, the original date, or any revised date duly fixed. The contractor may beat this date if he wishes and therefore be paid early, which the employer may or may not welcome doing, but not be paid a bonus without some special and non-standard provision being included in the contract. On the other hand, if he finishes late the next clause comes into play.

Clause 23.3.1 confirms that the contractor retains possession of the site until practical completion occurs and that insurance, by whichever party, continues until then, while the employer is not to take possession until that time. The rest of clause 23.3 provides for the possibility that the employer may 'use or occupy the site or the Works or', more likely, 'part thereof'. This may be done only with the contractor's consent and is a purely temporary arrangement followed by repossession by the contractor. It is also provided that the insurers are to agree, with the possibility of an additional premium being met by the employer.

JCT clause 24(2.5) – Damages for non-completion

The trigger under clause 24.1 for damages arising is the contractor failing 'to complete the construction' by the completion date, that is by the contract date or any extension of it already granted. The employer must, however, give the contractor notice that he has not completed. This presumably acts as a reminder to the contractor of the date and to seek any further extension that he can justify. He may even contend that he *has* finished. This is one of those critical instances where the parties are affected by the absence of an architect for contract purposes and where the evidence disappears with the passage of time.

The procedure under clause 24.2.1 is then for the employer to deduct liquidated damages from monies due or recover them as a debt. These are given in the appendix and are usually stated as an amount for each week that completion is delayed. They cease to accrue with practical completion under clause 16 and their weekly or other rate will be reduced if there is partial early possession by the employer under clause 17 or other similar arrangements, as discussed in Chapter 10.

The principles of liquidated damages are long established, but key points may be given here. They may be defined as a reasonable pre-estimate of loss due to a specific cause, here what the employer will suffer on late completion of the works, agreed between the parties and stated in the contract. They are not adjustable if the loss turns out to be more or less, but the courts will set them aside (despite the prior agreement of the parties) if they have been set at a level so high as to be punitive. Unliquidated damages based on the actual loss will still be available to the employer, if justified.

The contractor does not have to start paying or allowing damages on receipt of the employer's notice under clause 24.1, but only when the employer requires him to do so. This may be at any time up to final settlement as defined (see Ch. 10), after which the employer loses his right to require damages not already set out in writing. Any outstanding may be deducted from the final balance before payment.

It may be that the contractor receives a further extension of time which reduces his liability, after he has already paid over damages. If so, clause 24.2.2 provides for the employer to reimburse the excess.

JCT clause 25(2.3 and 2.4) – Extension of time

The prime purpose of this clause is twofold in allowing the completion date to be made later for defined reasons. It gives relief to the contractor so that he need not price his tender to cover the risk of having to meet damages for delay in completion due to these particular causes. But among the clauses are some which are the fault of the employer. If they were not included and one of them happened, the contractor could set aside the completion date without incurring liquidated damages, so long as he completed within a reasonable time for the whole works, which might be longer than a tight contract period extended only by a reasonable time for the delay element. The clause is therefore of potential help to both parties. Its list of causes, called 'relevant events', is as extensive as those in other JCT forms and very similar.

Procedures over extensions

Clauses 25.1 to 25.3 contain a set of procedures which are critical for granting an extension, on the basis of the events in clause 25.4. These procedures are those of other JCT forms, except that the employer plays the part there taken by the architect in giving extensions of time (referred to as 'fixing a later date as the Completion Date'). This is an area in which an employer acting without the advice of a consultant can easily be particularly at risk, in that strict observance of the procedures is essential to not losing a defined and disciplined completion date, while the fixing of a revised date usually requires a technical understanding of the issues.

The stages of the procedures in outline are as follows:

(a) The contractor gives notice 'forthwith' when delay occurs or is *likely*, with the cause and the relevant event identified.
(b) With the notice or as soon as possible, he gives the *expected* effects and an estimate of the *expected* delay.
(c) Thereafter he brings this information progressively up to date.
(d) The employer fixes a new and later date within 12 weeks of receiving adequate information, but does so not later than the completion date so being fixed, subject still to information. If it is not reasonable to fix a date within the time-scale, the employer is to notify the contractor and presumably fix a date as soon as he then can.
(e) Stages (a) to (d) are repeated for each delay that arises.
(f) After the first round of stages (a) to (d), the employer may take account of work omitted from the contract since granting the last extension, either by reducing the next extension granted or by fixing an earlier date in isolation.
(g) At any time after the completion date has passed, if earlier than practical completion, the employer may fix a final completion date and must do this at the latest within 12 weeks of practical completion. The date may be fixed as earlier, later or the same as that already applying, by reviewing previous decisions or taking further relevant events (not necessarily notified by the contractor) into account.

Unless the contractor acts in the first place he is not entitled to an extension, except by the employer's review of events after completion. But when once the contractor has acted and supplied adequate information, and also after completion, the employer must act or he will lose the benefit of a closely defined finishing date. The contractor would then be able to finish within a reasonable time, and not be liable for damages unless and until that time was held to have expired, or assert that he had done so.

The wording of the procedures omits, in reference to the employer, the wording 'as he then estimates' and 'in his opinion' which qualify the actions of the architect under the JCT standard forms. This emphasises that the parties are working face to face, whatever advice they may be receiving, without any intermediary. This form and the others all allow arbitration over clause 25 to proceed before practical completion. If a gross difference between the parties here leaves severe doubt about the completion date, it can therefore be resolved. But arbitration is usually better avoided, as the date is still open to review after completion when the element of forecasting has gone. Even an arbitrator cannot make certainties out of uncertainties during progress, and cannot remove the employer's right of review (subject to further arbitration) after completion.

Two riders to the procedures are given as in the JCT standard forms. First, the contractor must act to mitigate the effect of any cause of delay; he will on general legal principles be responsible for any delay that he might have avoided. This rider usually applies solely to such actions as progressing deliveries and adjusting the programme of work, where the contractor has the additional incentive of keeping his time-related overheads in check. In this contract, there is the further possibility of design or specification modifications at the contractor's discretion and without financial adjustment within the terms of the contract, or more drastically by a change in the employer's requirements under clause 12 leading to financial adjustment. In relation to most of the relevant events, there is no reason why the contractor should take such action simply to counter delay and in the absence of a properly instructed change. He may choose to respond to the employer's proddings, the more so if a supply-related relevant event is in question; but this will usually be more than what 'may reasonably be required' without payment. Clause 25.4.10 is an example of when the contractor may need to go further, as is suggested in comment upon it. A related issue of extra design with reimbursement arises in respect of disturbance of progress leading to loss and expense payment in clause 26, under which it is discussed.

The second rider is that the employer cannot fix a completion date at any point in proceedings that is earlier than the original contract date, even though the contractor may choose to finish earlier, so that the effect of omissions from the works is limited in this respect.

Events leading to extensions

Most of the relevant events have identical effects to those in other JCT forms, and so comment here is selective over special features. Several of them correspond to 'matters' in clause 26.2 leading to loss and expense payment – see Table 8.1.

Force majeure (clause 25.4.1). This covers extreme events only.

Exceptionally adverse weather (clause 25.4.2). This covers various options, but the first word is significant.

Loss or damage by the specified perils (clause 25.4.3). In itself this is covered by insurance. The parties may each wish to insure further to cover the costs to them of the delay, as distinct from the reinstatement and disturbance.

Civil commotion etc (clause 25.4.4). The results are not limited to effects at site alone.

Particular employer's instructions (clause 25.4.5). The clause references cover resolution of divergences over the site boundary, changes in the employer's instruc-

Table 8.1 Relevant events and related matters under JCT with design form

Cause of delay etc	Relief or redress available to contractor			
	Extension under clause 25	*Payment under clause 26*	*Determination under clause 28*[1]	*Determination under clause 28A*
Exceptional weather	Yes	No	No	No
Strikes etc	Yes	No	No	No
Delay by statutory bodies	Yes	No	No	No
Inability to obtain labour and materials	Yes	No	No	No
Exercise of statutory powers	Yes	No	No	No
Instructions due to antiquities	Yes	No[2]	No	No
Force majeure	Yes	No	No	Yes
Specified perils	Yes	No	No	Yes[3]
Civil commotion	Yes	No	No	Yes
Terrorism	Yes	No	No	Yes
Opening up and testing	Yes	Yes	No	No
Provisional sums	Yes	Yes	No	No
Lack of ingress or egress	Yes	Yes	Yes	No
Delay in possession	Yes	Yes	Yes	No
Delay in instructions, decisions etc	Yes	Yes	Yes	No
Delay by direct work	Yes	Yes	Yes	No
Employer's changes	Yes	Yes	Yes	Yes/No[4]
Divergences due to employer	Yes	Yes	Yes	Yes/No[4]
Postponement by employer	Yes	Yes	Yes	Yes/No[4]
Hostilities	No	No	No	Yes

[1] Clause 28 contains other grounds for determination besides those which may also lead to extension. See also note 3.
[2] Clause 34 (Antiquities) makes its own provision for the contractor to recover losses not otherwise provided for in the contract.
[3] Determination following the specified perils is also partially covered in clause 22C.
[4] Determination is provided only in limited instances due to a statutory body's default etc.

tions, expenditure of provisional sums, postponement of work and dealing with antiquities. Other discrepancies or divergences are not allowed, as they are in the JCT standard forms, because either the contractor's proposals are held to prevail by the third recital of the articles or clause 2.4.1, or because discrepancies are solely within the contractor's proposals and so his responsibility, subject to the next relevant event. As with other relevant events, none of these necessarily have effects leading to an extension. In the further case of opening up work or testing, any delay is borne by the party who also bears the cost of the action.

Delay in receiving any employer's instructions, consents etc (clause 25.4.6). Here the simple delay may be a relevant event, irrespective of the content. The scope is very wide, but is limited to those instructions and so forth 'which the Employer is obliged to provide or give under the Conditions', so that instructions over such matters as changes, testing and postponement are not included, as they are at his option. The decision of the employer over discrepancies in the contractor's proposals is covered specially by the clause reference, as it leads back to an error on the part of the contractor, who does not receive an extension for the inevitable delay that results. Nevertheless, the employer must act promptly once the contractor has made his proposal under clause 2.4.2, but not before, and cannot leave the solution to the contractor.

The proviso given is at best so positioned as to render the clause ambiguous. It is doubtless meant to apply to all of the employer's actions listed (so paralleling the other JCT forms), but could be read as applying only to 'including a decision under clause 2.4.2'. These words should be in parentheses to give clarity. The purpose of the proviso is to exclude cases in which the contractor asks the employer to act so late that delay is inevitable, or so far ahead that delay is not caused, but may appear to be caused. There are virtues all the same in the contractor listing, as soon as possible and in as few lists as possible, all the detailed actions which he requires of the employer, provided that he gives dates in each case (the actions required of the employer in principle under the contract are listed in Table 7.1). This is particularly true when design responsibility and knowledge of the extent of work and its programming lie with the contractor.

Delay in receiving statutory permission or approval (clause 25.4.7). This is an additional clause under this contract and takes the place of the rather similarly worded clause about nominated firms under other forms. While the employer may have obtained various permissions and approvals and have said so in the employer's requirements, all those that are outstanding are the responsibility of the contractor under clause 6.1.1, unless the employer has reserved any to himself in his requirements. Subject to this exception, the contractor carries any design risks and cost. The present clause means that the risk of delay does *not* also pass to him, but remains with the employer as in the other forms. This is subject to the contractor seeking to minimise any delay and so, again, to mitigate his loss. Pursuit on his part of an unreasonable design solution in the face of problems over regulations, for instance, would not therefore lead to an extension.

There are two further causes of delay over statutory matters, other than initial lateness, and neither count under the present relevant event. The first is divergence between them and the employer's requirements or contractor's proposals, dealt with in clause 6.1.2 by requiring the contractor to deal with his oversight at his own expense. This principle would run through here, although, if further 'permission or approval' were required, the contractor would be entitled to any extension due to its delay, unless his oversight be held to constitute the omission of a 'practical step'. The second cause is an amendment due to statutory requirements communicated after the date of tender, as envisaged in clause 6.3. This falls under the last relevant event in the present clause.

Work or supply of materials by the employer etc (clause 25.4.8). This clause reads oddly by making any work or supply count, even though it may simply be what was always intended: 'failure' in each case is more intelligible. The employer does have the power to introduce unexpected persons with the contractor's consent by clause 29.2 and this may well lead to delay. In all cases, including the first, delay must be demonstrated so that no real difficulty arises in principle.

Statutory powers affecting labour, materials or energy (clause 25.4.9). These must be exercised after the base date and by the United Kingdom Government, rather than by such bodies as national corporations or the police on their own initiative. Otherwise, this is a wide provision, overlapping to an extent with the next.

Inability to secure labour, goods or materials (clause 25.4.10). The key expressions here, as in all JCT forms, are 'inability', 'beyond his control' and 'could not reasonably have foreseen'. These mean far more than shortage except at a price. According to how advanced the design was at the base date, the clause allows for some variation in interpretation: the nearer the design to completing, the nearer the clause to standard interpretation. With a relatively sketchy design, it may be unreasonable to expect too much foreseeing by the contractor. Alternatively, it may be unreasonable for him to develop his design in directions prone to shortages. It is not practical to enunciate many general rules. The comments about clause 25.3.4 are relevant here though; the contractor may lose an extension if he fails to avoid a delay which he has virtually built into the contract by his own imprudence. This consideration overlaps with clause 8.1 about what is procurable, so that amended design or specification may be forced on the contractor, subject to the employer's consent.

Work by a local authority etc (clause 25.4.11). This provision relates to 'work in pursuance of its statutory obligations', the cost of which may or may not be included in the contract sum, and not to work as a sub-contractor. Only the former can lead to an extension. There is no rider about the contractor taking 'steps to avoid or reduce' delay, which is perhaps recognising the inevitable! Where work is predictable, the contractor should accommodate it in his programme, so that he is using 'his best endeavours' to prevent delay.

Failure by the employer to give ingress or egress (clause 25.4.12). Such ingress or egress is through land other than the site and may be for some part of the contract period only. It may have been set out in the employer's requirements and is then subject to any specified notice from the contractor. It may be 'as otherwise agreed' which suggests *post-contract* agreement, so that the measure of extension of the *contract* period becomes difficult to determine. There is no mention of the contractor's proposals which could well contain the particulars, especially in amplification of the employer's requirements. This could be sneaked under the umbrella of 'as otherwise agreed'.

Change etc in statutory requirements (clause 25.4.13). This event is entirely extra and needs to be made explicit because the contractor is performing the design. In other forms, it lies behind architect's instructions and need not be mentioned. It is referenced to clause 6.3.1 covering a change in the basic statutory requirements of all types, and to clause 6.3.2 covering an amendment to the contractor's proposals, due to a permission or approval under development control requirements only. In both cases, only amendments or changes after the base date qualify. No distinction is made here, as it is in clause 6.3.2, between amendments which introduce a deemed change in the employer's requirements, so that there is an overlap with clause 25.4.5, and those which the contractor must carry without adjustment of the contract sum. All such amendments may constitute a relevant event.

Deferment of giving possession of the site (clause 25.4.14). This is an optional right of limited deferment open to the employer by clause 23.1.2. The extension resulting will depend on its effects and need not be of the same duration as the deferment itself.

Terrorism (clause 25.4.15). This covers actual or threatened terrorism and official activities in response to either.

Loss and expense

JCT clause 26(4.7) – Loss and expense caused by matters affecting regular progress of the works

The purpose of this clause is to allow the contractor to be paid for particular disturbance effects leading to loss and expense. It is often referred to as 'the claims clause', somewhat loosely, and follows other JCT versions so far as possible. It is shorter in its opening part because there is no elaboration of procedures, the architect and quantity surveyor not being mentioned. The effect is likely to be much the same, as the precise procedure is often changed.

Procedures over loss and expense
There are, however, several critical features in clause 26.1 which must be observed (see the present author's *Building Contract Disputes* for details).

(a) The contractor must write to the employer to start activities, and must do so as soon as he is aware that a situation has arisen or is likely to arise. Although this is termed an 'application', it is a notice of trouble and not a statement of any sum of money. If he does nothing he may prejudice his right to payment, or at least render it difficult to keep records etc as its basis. The employer may then be justified in disputing the amount.
(b) A situation for this purpose arises when 'regular progress of the Works' in

whole or part is affected 'materially', or this is threatened. Disturbance of regular progress may or may not delay the programme and so also lead to extension of time: there is no necessary connection. The key feature is that work becomes dislocated or out of sequence.

(c) The contractor is to be reimbursed for his *direct* loss and expense. This reflects the legal concept that some consequences are too remote to be taken into account. Drops in productivity, additional supervision and other immediate extra expenditure on the works are all allowable. So too are costs of any prolongation that there may be, such as extra charges for temporary facilities. Some profit, financing charges and inflation amounts may also be taken into consideration in suitable cases. Effects in other contracts in general are indirect or remote. While these are the categories of reimbursement, the appropriate level of reimbursement will depend upon which costs fall within the fluctuations provisions applicable in the contract. Costs that do shall be assessed at the base date level and others at the current level. The distinction is drawn out under clause 38.
(d) Payment is to be made only if no other provision in the contract deals with it.
(e) There is a limited list of causes leading to admissible loss and expense, given in clause 26.2 considered below.
(f) The contractor is to provide the employer with reasonable 'information and details', to support his contention of disturbance and also the amount of loss and expense.

The above procedure is to be followed each time a distinct cause arises, even if its effects overlap with a preceding cause. If a cause goes on festering away, it is sufficient for the contractor to keep the employer aware of this and provide further data. Whereas the other JCT forms require the architect or quantity surveyor 'to ascertain the amount of . . . loss and/or expense', this clause leaves it between the parties on how to proceed. The inference is that the contractor will state the amount for the employer to check. This is what usually happens and modifies the printed procedure under the other contracts: for 'information and details' read 'claim document'. Where possible, progressive agreement is desirable. As indicated, the contractor risks not securing proper recovery if he holds everything over until the end of work on site, even though precise observation of the clause is not a strict condition precedent to payment.

Matters leading to loss and expense

Clause 26.1 names deferment of possession of the site as a 'matter' or cause leading to reimbursable loss and expense, while clause 26.2 gives seven further matters, one of which is sub-divided, six are among the larger number of relevant events in clause 25 leading to extension of time, although they are arranged differently (see Table 8.1). These six all stem from the employer's action or inaction, and in these matters he has to be on his toes and cannot leave the whole 'package' to the contractor. Of them, only that in clause 26.2.6 need be mentioned further. Here it is the disturbance effects of employer's instructions over changes and provisional sums that are in view. The valuation of the changes etc themselves, executed in an orderly manner even if out of sequence, is covered by clause 12. When instructions change an obligation or restriction imposed in the employer's requirements, it may become very difficult in practice to maintain any separation, conceptually or financially.

The distinct matter is given in clause 26.2.2, although it is really a narrower aspect of what is in clause 25.4.7 for extension of time. That clause refers to delay in permissions and approvals, 'of any statutory body', while this clause is restricted to delays of such items 'for the purposes of Development Control Requirements', that is questions of planning. These may well have their roots in what the employer cleared initially or what he has laid down fairly precisely in the employer's requirements, so that the contractor's room to manœuvre is somewhat restricted. They are also quite fundamental aspects. Whether the employer or the contractor has initiated any application, once the contractor has taken over post-contractually he is to be reimbursed for disturbance due to delay. Less fundamental matters, such as delay related to building regulations, are left at the contractor's risk so far as expense is concerned, although this is not the position over extension of time. There is also no reimbursement of loss and expense for disturbance due to changes in any statutory requirements, although again extension of time may be available. Here the contractor is far more likely to have set up the situation by his own choice, even if he could not have anticipated the changes. This restriction of cause for reimbursement is not unreasonable in the way that it distributes risk, although other lines could be drawn.

Clause 26 does not make any statement about the contractor mitigating his loss, as clause 25 does about his delay, but on general legal grounds he should do so and not let things go from bad to worse. The nearest reference is that in clause 26.2.2 to him taking 'all practicable steps to avoid or reduce' planning delay. This relates to a potential cause of loss, rather than to the consequential loss itself, and is the one cause over which he may be able to act, once it is threatening. In two other matters, giving notice about ingress and egress in clause 26.2.5 and applying for *necessary* instructions etc in clause 26.2.7, the contractor must act first to trigger the employer's action which only then may fail. Failure at any of these points will lose the contractor his right to reimbursement, at least in part. Mitigation, however, relates more directly to assuaging the consequences when once the cause is unstoppable.

The consequence is 'regular progress ... materially affected'. This may occur on the site or in workshops, as in other types of contract. It may also happen in this case in the design office, where too the contractor may suffer loss and expense and must seek to mitigate them. In all of these cases he must therefore rephase activities, delay or hasten the build-up of resources and take all reasonable needed action in relation to the works as he has already planned them, both in design and execution. In none of these cases may the employer instruct the contractor what to do within his own organisation, although the employer is entitled to decline to pay for costs that the contractor has not reasonably abated, providing he has been warning the contractor *en route*.

There is the further question of whether the contractor redesigns any work to help matters, as may occasionally be feasible, perhaps to introduce a different sequence of site work. He must not in any circumstances redesign them on his own initiative, in such a way as to introduce a change within the meaning of clause 12, that is, to the employer's requirements. If his design is still fluid, he may be able to finalise it differently to reduce his loss in some such way. Even if he has completed some element of design, or even built a part of it in the extreme case, he may be able to achieve a net saving by redesigning within the employer's requirements, as he always may within this contractual framework.

In any of these cases of design or redesign, the contractor should carry the employer with him, to observe clause 8.1 as well as the present. If he seeks to introduce a change, he must be instructed and then payment will follow. But in the other cases, when the design option is his, there are the related costs of possible extra design work and a presumably dearer construction solution than he intended. He should therefore obtain the employer's written consent or refusal to all that is involved, so that he is covered for reimbursement. Otherwise he may be reimbursed only for the actual disturbance costs themselves. He cannot claim the larger amount that would have accrued if he had not taken action: clause 26.1 allows only 'the amount . . . which has been or is being incurred'. But he should not delay work waiting for the employer's response; if the employer does not react, the contractor should carry on with work as already designed.

With all this said, there is no express or implied obligation upon the contractor to do anything but continue with the intended works as efficiently as is practicable in the circumstances of disturbance, as he would under other types of contract. Even if the employer himself suggests some design amendment, he can enforce it only by instructing a change. It is in some ways as well that, in the circumstances of disturbance, design amendment is usually too ineffectual or too slow to be worth considering!

Subsidiary provisions

Two short provisions round matters off. Clause 26.3 covers adjustment of the contract sum and, via clause 3, leads to interim payments on amounts. These by clause 30.2A.2 or clause 30.2B.2, as may apply, are without deduction of retention. 'Any amount from time to time ascertained' allows approximate or part amounts on account to be included.

Clause 26.4 preserves for the contractor 'any [of his] other rights and remedies'. While clause 26 is the last payment provision in the contract to be called upon (so clause 26.1), beyond it lies the possibility of arbitration or court action in the event of dissatisfaction. The contract may thus end in a whimper or a bang: which may be a matter for obscurity or for a law report!

Injury, indemnity and insurance

The three 'Is' are represented by three clauses which are almost identical with those in other JCT contracts and which therefore have mainly identical effects.

The differences in wording and effect all arise out of one issue: design. All other issues are simply summarised here, but should be pursued elsewhere in the literature or advice sought in this specialist and important area.

JCT clause 20(6.1 and 6.2) – Injury to persons and property and employer's indemnity

This clause *is* identical with its counterparts. It is a 'one way' provision, in that it is only the contractor who indemnifies the employer. Its first two parts deal respectively with injury to persons and to property, and they draw the line of the contractor's responsibility differently:

(a) Over persons, the contractor is liable unless there is 'any act or neglect of the Employer', or of others for whom the employer is responsible. The contractor thus has to demonstrate that the employer is liable, to be free himself.
(b) Over property, the contractor is liable only when there is 'any negligence omission or default' on his part, or that of others for whom the contractor is responsible. The position is reversed, so that the employer has to demonstrate the contractor's liability.

The effects are that the contractor is liable under (a) and the employer under (b), if no negligence can be shown. Clause 20.2 excludes from the contractor's liability any existing structures and contents which the employer is responsible to insure under clause 22.C.1. It also states that it is subject to clause 20.3, which defines 'property real or personal' as excluding the works. This removes the works entirely from the present clause and so removes any potential overlap with whichever part of clause 22 applies, so that all insurance of the works must be traced in that clause.

This clause has no obvious bearing upon design, as it applies to injury from 'the carrying out of the Works', as distinct from defects in its construction when complete, or arising from a design fault during or after construction. This may of course be a difficult distinction to draw. Both article 1 and clause 2.1 appear to differentiate construction from the design leading up to it, which is the contractor's responsibility under clause 2.5 and over which he should insure, even though this contract does not require it (see Ch. 5). Nevertheless the clause does apply to a designer visiting the site, who may still run his car into another vehicle or drop his briefcase on someone's toe.

JCT clause 21(6.3) – Insurance against injury to persons and property

The indemnities of clause 20 are to be backed up by insurances taken out by the contractor and his sub-contractors to protect him and the employer, against the possibility that their resources are too slender to meet a claim. Clause 21.1.1.1 therefore repeats wording from clause 20 to give the terms of insurances, while its opening words emphasise that the insurance is as well as and not instead of the contractor's indemnity. Clause 21.1.1.2 affirms the contractor's statutory responsibility to insure as an 'employer', while referring to appendix 1 for the minimum sum to be covered over his other insurance against his indemnity. Because the employer stands to be at risk if the contractor does not insure in this way, he may call for evidence of the insurances under clause 21.1.2. He may himself insure under clause 21.1.3 if there is a default, and deduct from payments or otherwise recover the premiums. All this is as in other JCT forms.

Clause 21.2 deals with a special insurance against damage to 'any property other than the Works', which is bound to be near by and which may or may not belong to the employer, on broadly similar lines to other forms. It is against a limited range of risks 'arising out of . . . the carrying out of the Works', when the contractor is not negligent in performance and which all lead to structural damage. Unlike the preceding insurances, it is to be in the joint names of the employer and contractor, because there is the possibility that a claim could be brought against either or both of them. The sums covered are to be given in the employer's requirements, as well as entered in the appendix this time, because the contractor needs to include in his tender and take out a special and expensive insurance, not covered by his running policy. Indeed, unless the insurance is mentioned in the employer's requirements, the contractor does not have to take it out at all, as it is needed only in particular circumstances. To avoid doubt, the employer's requirements should therefore state if it is *not* required. If, however, the need is present, the employer could be at great risk without the insurance.

The major difference in the wording is the omission of any reference to a provisional sum to cover the costs. Instead, the contractor must include in his tender an amount which is not subject to any adjustment later. Under other contracts, the provisional sum method is used because the contractor may not be able to assess the precise physical risks, arising out of the design and site with which he is not deeply familiar when tendering. Under this contract, the contractor deals with the design and must assess the need to support adjoining structures when tendering. He thus has the information for obtaining an insurance quotation.

The rest of the wording of clause 21.2 is as given in other versions, but some comments arise for that very reason. The causes of damage exclude:

(a) Negligence etc of the contractor etc: this relates to the activity of construction. If the contractor or others are negligent, their liability is under clause 20, backed by the insurances of clause 21.
(b) Errors or omissions in design: these are covered by the contractor's liability under clause 2.5 and any insurances that he maintains.
(c) What is reasonably seen as inevitable: this is hardly insurable! It is presumably the contractor and his insurer who see the inevitability, as the employer may be in no position to judge. If such a contingency is known, the contractor should draw the employer's attention to it in the proposals. Otherwise he may be liable for not disclosing the full import of an imprudent design.
(d) Clause 22 perils which are at the employer's risk: clause 22C is the major case

here, with existing structures and contents involved. Under both clauses 22B and 22C, materials on site are covered and distinguished from the works.

(e) Nuclear risks etc: these are common exclusions.

Of these exclusions, (a) and (b) relate to the default of the contractor. The insurance therefore covers damage which occurs when the contractor performs his activities properly, when the damage is not inevitable and yet occurs. This is a limited but important sector. Whether exclusions (a) and (b) apply under this policy and then are insured against elsewhere, is a matter for the contractor; the employer is not directly concerned with the distinction.

Under clause 21.2 the insurers are to be approved by the employer, as in other forms. But, whereas in those forms this relates to a choice of insurers against an adjustable sum, in this form the contractor has to choose in relation to the premium before entering into the contract. He should therefore include sufficient information for the employer's approval in the contractor's proposals, so that he has clearance before commitment. Again, if the contractor fails to insure, the employer may do so and recover. As the employer could face a direct claim, or one for subrogation, he should certainly act in this instance.

Clause 21.3 is the usual provision over the excepted risks of nuclear perils etc which are either covered statutorily or are uninsurable.

JCT clause 22(6.6 and 6.7) – Insurance of the works

This very long clause consists of five sections:

Clause 22: General clause introducing the remainder.
Clause 22A: New works insured by the contractor.
Clause 22B: New works insured by the employer.
Clause 22C: Works of alteration or extension with existing structures and contents, all insured by the employer.
Clause 22D: Employer's loss of liquidated damages insured by the contractor.

Of these, the middle three are alternatives to one another and usually only one will be in use, while the last is optional. They follow closely on the versions found in the JCT standard forms and largely need only summarising.

Clause 22 defines all-risks insurance, which in insurance practice is never precisely what it may appear to be. It relates to insurance of the works and site materials (themselves also defined) and a number of exclusions from its scope are given. The clause ends with a provision for a term in the policy to protect sub-contractors from the insurer seeking contribution when there is a claim, either by recognition or waiver.

Clauses 22A.1 with 22A.2 and 22A.3 (together), 22B.1 and 22C.1 and following clauses deal with the insurance in common ways as follows:

(a) The value insured is 'full reinstatement value' and not the amounts in the contract sum analysis.

(b) The contract obligations run up to practical completion or determination, after which the employer is entirely responsible.
(c) The specified perils are set out in clause 1 and cover the major accidental and natural hazards, even though the contractor may be negligent, although such matters as theft and vandalism are not specified. The contractor is advised to cover unnamed risks in his own policy in each case.
(d) Cover is for the works, site materials and professional fees and in one instance existing structures and contents, all as noted above. Temporary buildings, plant etc are excluded. Works and materials of sub-contractors are covered, so that they need not insure these elements.
(e) The policy is to be in joint names, so protecting the employer against the contractor's insolvency, but also either party against subrogation if he has been negligent.
(f) The insurer is to be approved (except when clause 22A.3 applies), while if one party fails to insure the other may do so and recover, as under clause 21.

Clauses 22A.4, 22B.3 and 22C.4 deal with the results of damage in fairly common ways as follows:

(a) The contractor is to clear debris, restore the works and proceed to completion. An extension of time is likely to be granted under clause 25. He is not obliged to reinstate any existing structures, although he may well agree to do so.
(b) The contractor is to be paid for the extra work with its disturbance costs. In clause 22A.4, the payments are by the insurers and in the other clauses by the employer (who recovers as he may from his insurers) as though a change had been instructed.
(c) In all cases the contractor receives the money by interim payments, without deduction of retention by clause 30A.2.2 or 30B.2.2.

Apart from procedural points, the main difference under these clauses is an optional right available to either party in clause 22C.2 to determine the contractor's employment. This is subject to a right of arbitration open to the other party. The grounds are 'if it is just and equitable'. Likely grounds are the impossibility of reinstating the works without first reinstating the existing structures, which in scale or character may be beyond the contractor's resources, or that the employer may not wish to reinstate, but to start again, even on a different site. The pattern of settlement after determination is that of clause 28.2, but without payment to the contractor for his loss due to the determination.

The differences in the foregoing scheme in the present contract are few. Clauses 22A.1 and 22A.3.1 add to 'the full reinstatement value' the term 'together with the costs of the design work of the Contractor', while the usual insertion of a percentage in the appendix for 'any professional fees' is qualified as 'incurred by the Employer'. While the employer has to make an uncertain stab at the very variable amount which he may incur, as usual, the contractor has to cover the costs in design related to full reinstatement. These may be comparatively limited and extend principally to a survey of what survives, supplementary drawings etc only and supervision during reinstatement. Most of the original design should be suitable to use again. If only partial reinstatement is required, some of these elements may be proportionately higher.

A more complex situation may come about if it is expedient for those concerned to adapt the design in the light of what remains, perhaps to regain as much lost time as possible. The contractor and his insurers may thus settle on a rather different distribution of design and construction costs. The employer is still entitled to the 'same' building as before: he should therefore regard any proposed amendments of design in the same light as any which the contractor might put forward during the natural flow of the works, that is as subject to his 'consent', while not leading to an adjustment of the contract sum (see general discussion in Ch. 5). If the employer wishes to introduce any amendments for his own ends at this juncture, he must instruct a change and pay the difference. How this affects the insurance settlement is none of his concern.

Clauses 22B and 22C make no mention of design or other fees, just as in other JCT versions, so that the rule in clause 12.4 about allowing for design covers the point. All restoration and other work are to be paid for by the employer 'as if they were a Change in the Employer's Requirements'. Any amendment to which the employer consents, and any changes proper which he instructs will automatically be taken up in the payments made. These will allow for piecemeal working and all other factors and are likely to include a high proportion of daywork. They are also likely to have only a distant relationship to the contract sum analysis. The employer must negotiate his own settlement direct with his insurers, although he should be able to pass amounts on. In essence, this process is the same as under clause 22A, with the relationship of employer, contractor and insurers re-arranged. Only if the contractor seeks to introduce expensive amendments to remedy defects in his own design work, will he be in conflict with the other persons. If the employer is in the middle, or if he has no insurers behind him, he will need to look closely at any such issues over division of costs.

Under any of clauses 22A, 22B or 22C the contractor faces a particular cash flow problem if stage payments under clause 30.2A apply, unless he makes special arrangements. All payments for clearance and reinstatement are due with other interim payments and, while periodic payments under clause 30.2B would continue at the usual intervals, stage payments will be retarded by the damage. Taking a simplistic example of damage which affects the whole of the works and takes three months to put right, the contractor will be outstanding on his expenditure in clearance and reinstatement (as well as any stage partly completed) during that time and until the next stage is reached under clause 30.3.1.1. When he insures, he should arrange that payments will be made earlier and that the employer is prepared to pass them on. When the employer is responsible for payments with or without insurance, the contractor should also make arrangements with him. In all cases he will need to embody provisions in the contractor's proposals to make these arrangements effective.

Clause 22D has no connection with the other sections of clause 22 and its organic connection with the whole contract is difficult to discern. It is optional and allows the employer to secure insurance through the contractor to protect himself against the loss of liquidated damages when the contractor secures an extension of time. It would seem more practical and direct for the employer himself to insure outside the contract should he wish for this cover. The clause is also of somewhat uncertain force in its detail.

Determination of the contractor's employment

This term is used in the body of the JCT clauses dealing with this topic, although plain 'Determination' is used in their headings. It refers to the premature ending of work by the contractor, with this ending final unless the parties come to some agreement over reinstatement, on terms with which the contract cannot be concerned. Each clause provides terms to be used to see matters through to a conclusion in the unfortunate circumstances in which it comes into use. Determination of employment thus does not signify the end, or determination, of the contract itself: if this were provided, all the contract terms would be abrogated, including those in the clauses concerned. But also invoking the terms of one of the clauses does not take the parties past some point of no return: either of them still has recourse to 'any other rights or remedies' about the reason for determination or about details of the ensuing settlement, at least in the circumstances of default.

Either party may determine the contractor's employment (referred to hereafter as 'determination' alone) according to events, and the terms vary accordingly. The clauses are extremely close to the JCT standard forms in structure and detail. They differ principally by the addition of elements about design, which also has some effect on other identical wording. The main contents of the clauses are:

(a) Determination by an aggrieved party upon:
  (i) default by the other party in specified respects;
  (ii) insolvency of the other party (with an option of possible reinstatement if the employer has determined);
  (iii) corruption by the contractor only.
(b) Determination by either party upon a specified 'neutral' occurrence.
(c) Activities to complete the works, if the employer determines when aggrieved.
(d) Decision by employer not to complete the works, but to clear the site, if he determines when aggrieved.
(e) Activities to clear the site, if the contractor determines when aggrieved, or either party determines on a 'neutral' occurrence.
(f) Financial settlement for the part of the works performed and over subsequent activities.

Of these, (a) and (b) are alternatives to one another, and so are (c), (d) and (e) to one another, while (f) varies in content according to which of the others apply. The details of the clauses are not pursued here where they do not differ from other versions, as they may be followed up elsewhere (for instance in the author's *Building Contracts: A Practical Guide*). An outline only of each clause is given to place more detailed comments over design in context.

Over determination by either party and any eventual completion of the works, there hangs the cloud of what happens to design liability. The outgoing contractor remains legally liable, as though determination has not occurred, for the design of any work which he has constructed and for any further design, which he has supplied and which is used without need for revision or development. Under clause 2.5.1, the contractor warrants his design in its own right and as additional to his liabilities for constructing that design. The value to the employer of this warranty is no greater than the continuance of the contractor's solvency. The employer may thus face two persons between whom liability is split: the old contractor and the new contractor or consultant. Alternatively, the first of these may be lost, as much as may happen with a free-standing consultant. It is suggested that it may be possible for the employer still to proceed against the old contractor's consultants, if he had any and with whom the employer had no contract, *in tort* on the grounds of their duty of care to the employer who has relied on their skill (see Chs 5 and 10). If this is so, it gives hope to employers, but a strong incentive to consultants to review the terms of their professional indemnity insurance and of any collateral warranties before they are imposed on them.

It may well be that completion of the works is not attempted on a design and build basis, because of these complications and those on site. If so, and if the contractor's design team was not in-house, the tidiest approach is for the employer to engage the same team as direct consultants to him, on the express basis that they become responsible to him for *all* design, before and after determination.

JCT clause 27(7.1 to 7.6) – Determination by employer

Reasons for determination

Three of the ways of default in clause 27.2 relate directly to poor performance, whether by failure over progress or by defective work. The fourth bears indirectly on the same concern by dealing with unauthorised sub-letting. In these cases there is a procedure of notice by the employer, waiting to see whether the contractor mends his ways and, if he does not or if he resumes default, determining. A resumption must be of the same category of default out of the four to bring about the instant determination which is then possible.

Design is mentioned only in clause 27.2.1.1 over complete suspension of the works. As clause 27.2.1.1 gives the two elements of 'design and construction' as 'one respect' leading to 'such default' under clause 27.2 as a whole, it follows that default with a notice and resumption for one element leads to the possibility of determination without further notice for default in the other. Complete suspension 'without reasonable cause' excludes situations in which the contractor is prevented by causes beyond his control. It is far easier to demonstrate that suspension has occurred in the case of construction, where there is a presumption (reinforced by clause 27.2.1.2) that the contractor will proceed continuously with work on site. In the case of design, work is far more able to proceed intermittently without holding up work on site or in workshop, depending on how much detail is finalised before construction begins. It may thus be possible to suspend design wholly for a while without threatening progress. The employer should therefore proceed with even more care over design suspension than over construction, to avoid acting 'unreasonably or vexatiously' and risking damages. Even with construction, the presumption of continuous work may not always stand in a contract in which it is not the whole of the contractor's activity.

These considerations may apply in the different, but related, situation of not proceeding properly under clause 27.2.1.2, where 'obligations' include design, as there is no express differentiation of the contractor's functions. Other JCT versions give 'the Works' here. Action over design here must be taken more cautiously still.

Clause 27.2.1.3 deals with non-removal of 'work, materials or goods not in accordance with this Contract', where the fault may go back to poor design or specification, rather than show solely in site performance. The comments about design, quality and the employer's consent in Chapter 5 indicate the need to draw distinctions carefully here. In all cases the fault must also be so serious that 'the Works are materially affected'. This suggests a fault that might lead to structural failure or a failure of the building to satisfy its purpose, so radical as would cause significant expense or inconvenience to put it right after occupation, even if completion is possible.

In the case of unauthorised sub-letting under clause 27.2.1.4, design is introduced directly by its separate treatment in clause 18.2.2. It is clearly important to the employer that the works are properly designed in all their parts and critical that they are so as a whole. But sub-letting is just as significant for the contractor: he not only is paid the same to whomever he may sub-let, but he also has the content of what he is to build at stake, again with no adjustment of payment if a different designer makes design changes inevitable in some way. The parties should not let themselves into this dilemma, and the comments under clause 18 should be noted, especially over prior consent to major sub-lettings. The question of what happens to all sub-contracts when there is a main contract determination is also mentioned there, and may be the governing factor over whether the employer should go for a determination when design is an issue.

The second main route to determination, the contractor's insolvency, is the subject of clause 27.3. This is an intricate area and further the clause may not always be enforceable. It presents no special features in its own right, although there may be particular practical problems about design continuity, where again the comments under clause 18 relate.

Clause 27.4 about corruption needs no comment here, except to note how extreme it is and how potentially unfair it may be to the contractor when he is unaware of the corruption.

Settlement after determination
Clause 27.6 takes up matters following a determination when the employer has the works completed. Those which are in other JCT forms may be summarised as follows:

(a) The employer may employ others to complete the works and make good defects, or by implication may do so himself. In so doing, he and they may use plant, materials etc already on the site free of charge. This is subject to the rights of plant hirers, materials suppliers etc to require payment. Any items that are superfluous are to be removed by the contractor, otherwise the employer may sell them and hold the net proceeds for the contractor.
(b) The contractor is to assign any supply and sub-contract agreements if required to the employer, who may be able to reassign them to any incoming contractor. The employer may pay any supplier or sub-contractor for materials or work before or after determination, so far as the contractor has not paid. If the employer has paid the contractor already for these elements, this will result in an immediate double payment, and so should be considered only for special reasons. These might be to induce some person to continue with work when it is not expected that the contractor will yet become insolvent, so preventing full recovery from him. All of this may apply for consultants. None of it applies at all for anyone if the contractor is insolvent. At any other time it is optional.
(c) The employer pays nothing further to the contractor until the works are complete. The elements in the settlement are what was paid to the contractor before determination (not any revaluation to what should have been paid), plus all costs of completion, plus the employer's direct loss and damage, and less what the contractor would have received had he completed.

The extra provision is in clause 27.6.1, apparently over drawings etc which the contractor was due to provide before practical completion as record, maintenance and operating information. The precise scope of the clause is not easy to derive from its grammar. 'The purposes referred to in clause 5.5' are certainly about as-built drawings and so on. The contractor has to provide for the contractor's retention two copies of all that is listed in clause 27.6.1, seemingly falling into two categories. The first is 'all such drawings (etc) as ... prepared or previously provided'. Leaving aside the subtle distinction, this means two copies of all that the employer has already had under clause 5.3, meaning all that the contractor has prepared. Not all of it may suit the purposes of clause 5.5. The second is 'drawings and information relating to the Works completed before determination'. This phrase relates uncertainly to what goes before, but adds nothing to it, as it creates a tautology by referring to a sub-set of the previous list; that is, what has actually been used. The expression 'drawings and information' is that used in clause 5.5, but that does not help its meaning here. There is no requirement for the contractor to prepare any fresh data here, merely extra copies of what already exists. In the circumstances, the employer must take what he can get, which is realistic. If it happens that someone else is to complete the design, this information is available for use for the works within the normal copyright position (see Ch. 5) and will be most useful by providing continuity, which is the most pressing concern.

This latter possibility is envisaged by the inclusion in clause 27.6.2 of 'the design and construction of'. If indeed the 'other persons' mentioned are actually the same consultants as before, under an assignment or a new agreement, so much the better for continuity and a rapid resumption of work.

Clause 27.7 is an alternative to clause 27.6 and deals with events if the employer decides not to proceed further with the works. It follows its equivalent in the JCT standard form and requires the employer to notify the contractor within six months of determination whether he intends to proceed to completion. Otherwise the contractor may require the employer to state his intentions. When it is established that the works are not proceeding, the contractor is to be paid the value of work he has performed, less what he has already been paid and less the employer's expenses and loss or damage incurred.

JCT clause 28(7.7 to 7.13) – Determination by contractor

Reasons for determination

There is one omission from the list of reasons for determination given in the JCT standard forms: interference with or obstruction of the issue of a certificate, since there are no certificates here. Otherwise, the reasons given are close to those in other versions, with one addition. One reason in clause 28.2.1 is failure of the employer to pay the contractor, subject to a notice procedure, 'the amount properly due . . . under clause 30.1'. Clause 30.1 requires the employer to pay 'in accordance with clauses 30.1 to 30.4', while clause 30.4 allows him to pay less than the amount for which the contractor has applied, so long as he pays what 'he considers to be properly due'. As clause 30 leaves what is 'properly due' as a matter for discussion between the parties then, provided the employer pays a not unreasonable amount as he sees it, the contractor cannot apply clause 28.2.1.1. Determination here is available only if the employer 'does not pay . . . [an] amount' *at all*. The contractor thus has an extremely powerful weapon in limited circumstances. When there is unduly reduced payment, or even when there is none, he may rather seek immediate arbitration via clause 30.3.5, as an option that keeps the contractual ship afloat.

The other reasons for determination in clause 28.2.2 all apply when they cause complete or nearly complete suspension for a continuous period inserted in appendix 1. Suggested periods occur in a footnote, varying from a minimum of one month, according to the reason. Unless there are insertions, nothing applies and the clauses become inoperable over these reasons. The opening part of the clause differs from other versions in two ways. It refers to suspension of 'the construction of the Works', and so excludes suspension of design alone which may not cause construction to stop. If it does stop construction, the clause bites.

It also includes 'after the Date of Possession' as defining the timing, and presumably start, of suspension. This perhaps aims to rule out undue postponement of possession, discussed under clause 23, although *construction* can hardly be suspended before the contractor has possession of the site.

The reasons also occur in clause 25 over extension of time and clause 26 over loss and expense, and they are discussed mainly under the former (see Table 8.1). Only those with special features are mentioned here.

Employer's instructions (clause 28.2.2.4). This list does not include provisional sum expenditure and antiquities, as given for extension of time. When the instructions are due to either negligence or default of the contractor, they are not admissible.

Work or supply of materials by the employer etc (clause 28.2.2.6). Here 'delay in' prefaces the wording and makes the clause read more precisely than its counterparts. It means that 'delay caused by or due to' these reasons does not lead to determination, so excluding *inter alia* the introduction of direct work.

Delay in receiving permission or approval over development control (clause 28.2.2.8). This is the more restricted wording that applies in clause 26. The recommended delay here is two months.

Settlement after determination

Clause 28.4 deals with all matters following a determination, other than any completion of the works which the employer may be able to arrange. The elements which are in the JCT standard forms may be summarised:

(a) The contractor is to clear everything that is his from the site and allow subcontractors to do likewise, subject to leaving materials paid for by the employer.
(b) The employer is to pay the contractor for all work performed or started, loss and expense during progress, materials left behind, costs of removal and loss and expense due to determination (this last includes loss of profit).

Several matters add to or modify the above. Clause 28.4.2 requires the contractor to provide copies of drawings etc and is identical with clause 27.6.1, under which there is discussion. Clauses 28.4.3.1 and 28.4.3.2 both have '(including design work)', qualifying the value of work for which the employer pays. This means that the contractor is paid directly for the design which the employer may use for the purpose of the works (but not otherwise for copyright reasons; see Ch. 5), and which is included in the clause 28.4.2 copy documents, rather than indirectly as when the employer determines.

JCT Clause 28A(7.14 to 7.16) – Determination by employer or contractor

Reasons for determination
The reasons in this clause are what may be termed 'neutral', in that they either cannot or are unlikely to arise from any lapse of either party. They all relate to suspension of the works and do not differ from those in the JCT standard form.

In the case of delay due to the specified perils in clause 28A.1.1.2, the recommended period of delay in the appendix is three months when clauses 22A or 22B apply, whereas only one month is given for the other causes of delay. This reflects the strong possibility of delay in this instance. Clause 22C has its own determination provisions, already discussed.

Settlement after determination
The pattern here follows that of clause 28 in most respects. Three differences may be noted:

(a) Half of the retention is payable within 28 days of the determination, rather than at settlement.
(b) The employer prepares the final account, rather than the contractor.
(c) The contractor's loss or damage due to the determination is reimbursed only when loss or damage to the works due to a specified peril arises from the employer's negligence or default.

Chapter 9

Changes, fluctuations and fiscal matters

Use of changes and provisional sums

Valuing changes and provisional sum expenditure

Market and tax fluctuations in costs

Fiscal matters

Design and build contracts as a class need to be lump sum contracts, unless some other such control system as a guaranteed maximum price is used, so that the contractor receives a fixed payment in return for producing the works which he designs. As stressed in earlier chapters, there is no question of any difference in payment to take up changes in the contract design or developments of it in detail, which are forced upon the contractor by impracticalities in his intentions or similar contingencies, any more than to deal with estimating inadequacies in the tender. Pre-contract design and estimating are two sides of the same coin and, to confound metaphors, the contractor must lie on the bed he has made, as must the employer when once he enters into the contract, even if he finds an error against him in the contract sum.

There are, however, several major causes which may lead to an adjustment of the contract sum and which are set out in most building contracts of the various types. Design and build contracts are no exception over the leading causes as taken in this chapter:

(a) Changes introduced into the works by the employer, in other contracts termed 'variations'.
(b) Substitution of actual amounts for provisional amounts in respect of work agreed to be uncertain at the tender stage.
(c) Fluctuations in production costs.
(d) Disturbance of operations caused by the employer or otherwise his concern.

All the causes of adjustment under the JCT form are as listed in Table 6.2. Those not discussed may be distinguished by the lack of rules for their calculation in the various clauses. There are other financial provisions not termed 'adjustments of the contract sum' and dealt with by other means. Of these, two are considered here:

(a) Statutory tax deduction.
(b) Value added tax payment.

The JCT clauses dealing with all the foregoing items are covered here. They are to be found in Section 4 and part of Section 5 of the anticipated section headed revised JCT contract (revised numbers are in brackets and see also comparison table in the Index of JCT clauses). They are all modelled in principle and largely in detail on those in other JCT forms, so that discussion here is selective. Several points of wider application may be taken first.

Use of changes and provisional sums

Discussion under this main heading outlines some major principles about the contractual position considered in Chapters 5 and 6, which affect design development and change in the post-contract phase. For simplicity here, 'design' is taken to include obligations etc that may be changed, and 'changes' are taken to include provisional sum expenditure (these are separated in detail under the next main heading). These principles are also basic to the detailed considerations of instructing and valuing changes, etc. set out under the next main heading, which should be read accordingly. They are rooted in the nature of the contract as 'entire', but modified to allow changes within limits to its subject matter, and in consequence adjustment of the contract sum (see Ch. 1, over entirety).

Design and changes

Changes are common to many other building contracts as well, as later discussion of the JCT clauses shows. They have a distinct flavour in the present type of contract, because of the locus of design responsibility.

(a) The employer sets out what he desires in his requirements, but does not design the works.
(b) The contractor interprets the employer's desires and embodies the results at tender stage in his proposals, which include design and statements of intent about any residue.
(c) The contractor develops any stated intentions into final design, post-contractually and possibly during construction.
(d) The employer may modify his requirements within reasonable limits, again post-contractually and during construction.

While (a), (b) and (c) are sequential, (c) and (d) are broadly contemporaneous. In fine detail therefore the employer, in modifying his requirements or 'instructing a change' in contractual terms, may change what may not yet have been fully designed. This is reminiscent of the countryman asked the way by a passing motorist, who replied 'If oi wur yew, oi wouldn't start frum 'ere'. Not only is the employer unsure what he is getting in detail until it is designed, but he is not sure what he is changing. This may be held to be only a minor irritation, except for the question of payment. Without any change, the employer has to decide whether the contractor's developed design represents a fair interpretation of both requirements and proposals, remembering that the latter overrides the former if there is tension. With a change that precedes developed design, the parties have to agree in some quantitative way what the 'weight' of the development would have been, so that it may be compared with the change as finally designed, and the value of the change deduced.

There is no formalised way of doing this: much will depend on the detail embodied in the proposals and in the supporting contract sum analysis. The vaguer the one is, the more useful is the detail in the other. The JCT contract is not alone in sensibly ignoring this question, it simply provides for evaluation 'consistent with . . . the Contract Sum Analysis', without assuming any stage of design development. The point is not raised again in this chapter, but should be foreseen when examining the contractor's tender. It is patently fatuous to design to the original proposals simply to give a firm point of departure for valuing the change; indeed this could be open to abuse by an unfair contractor. This suggests that if the employer is merely toying with a partial change to a developing design, but is waiting to see how it develops, then it is better not to communicate his thoughts! He has to weigh this against the cost of two elements of design, for which he pays, and possibly delay to and disturbance of progress as well.

This chapter, like this book generally, proceeds on the basis of a firm lump sum contract. In passing, it may be noted that a guaranteed maximum price contract needs special consideration over changes. While the cost varies almost automatically with the changes, the maximum price does not. An assessed adjustment of this price is needed, so that the cost is compared with the correct amount.

Acceptance of changes

The reasons for changes etc to the design are various:

(a) The employer may change his mind and instruct accordingly.
(b) The employer must instruct about work covered in his requirements, but not in the proposals, by provisional sums or quantities.
(c) Changes may be deemed to have been instructed for such reasons as changes in statutory requirements and the costs of reinstating work damaged in particular ways (see JCT clauses 22B and 22C).
(d) The contractor may develop his design along lines that effectively produce a change, perhaps due to the impossibility or excessive cost of the original proposals.

If (a) is unreasonable, the contractor may object, so obliging the employer to modify or abandon his idea. This is usually a contract provision and safeguards the contractor's design integrity and, incidentally, the employer's right of redress if it is deficient. The parties have little choice about (b) and (c): they must 'happen'. If the employer considers that (d) is occurring, for whatever reason, he may object and require the contractor either to revert to his original line of development or to produce a satisfactory alternative, according with the contract documents. If then the parties cannot agree, arbitration or other proceedings are to hand, unless the employer is prepared to accept the (presumably) less favoured solution of the contractor as a change, and to pay less for it. In this instance, there is again an uncertainty element, due to the earlier partial design.

Procedure for changes

The key feature when the employer instructs a change is that he does so to modify his requirements, that is, he sets out the *end* that he seeks and not the *means* of its attainment. This is a matter for the contractor as he amends the design for which he continues to be responsible. Provision of design by the employer in a change instruction may well reduce the contractor's responsibility and affect his potential liability, as set out in detail in Chapter 5.

When the contractor produces a design response to a change, it is reasonable that the employer should be allowed to consent that it 'appears to meet his requirements', as with the original contractor's proposals, even though he does not so approve it as to take over responsibility. Curiously, the JCT contract in particular in its main part has no procedure for this, only in the supplementary provisions for use with the BPF procurement system (see Ch. 14). It goes no further than requiring the contractor under clause 5.3 to supply to the employer copies of all documents which he 'prepares or uses for the purposes of the Works'. These appear to be solely for information and no time is given for their supply, so that it could be after the related construction. If there is a 'deemed' change, say in response to a change in statutory requirements, the employer is the more likely to be behind the game and unable to act in time.

There are at least two good reasons why the employer should have the opportunity to object to a design change. The first, which applies also to a design development, is that he may simply not agree with it as fulfilling his requirements. The second, which does not apply to a design development, is that the employer faces a difference in the amount payable under the contract. When introducing a design change, the contractor does not have the incentive to economise which he has when tendering or subsequently developing design included in the tender. He is due under the usual interpretation of the valuation rules to be paid for the work content at a level consistent with that for similar work in the contract sum analysis. He may therefore anticipate being paid for what he designs, be it more or less than his original design. If the employer considers it excessive, and so leading to too large an addition or too small a deduction, he can try two ploys. The more obvious is to object to the design, but this may draw the employer into insisting on a reduced standard and so away from having redress over a failure.

The less obvious ploy is to say that 'valuation' of the change implies paying for what it is worth to the employer. The weakness of this is that valuation has long been used of custom to mean 'measuring and pricing what is done' and so to mean paying something more like cost than value. Even though 'what is done' is determined here by the contractor, the same philosophy is intended to apply. Further, the intention to use this ploy should be voiced in advance of construction of the changed work, and so it still implies a criticism of the design. It may be a valid bargaining point if the changed construction is produced without the employer having had a chance to query it in advance.

Some suggestions

By way of summary of the last three sections, it may be suggested that the parties:

(a) Note the built-in uncertainty over design development, its carrying over into changes and so into their valuation.
(b) Distinguish the reasons leading to actual changes in design, and which of these qualify as changes under the contract.
(c) Seek to anticipate the likely incidence of changes and difficulties due to uncertainty and overlapping reasons for changes and, if the design is too vague by these criteria, consider its further development *before* entering into the contract, remembering that arbitration on these issues during progress is neither available nor desirable.
(d) Agree to a procedure, preferably contained in the employer's requirements, for giving the employer early warning of the detail of all design development and design changes, so that he may check whether his requirements appear to be met.
(e) Accept that more tolerance and give and take, and less precision than apparently exists under other contract arrangements, may have to be displayed in dealings.

These suggestions should be balanced against the pressures of other practical issues, and they are more important for some projects than others. An employer or contractor new to design and build could, however, run into unexpected difficulties if faced with an experienced and ruthless opposite number. On the whole, some expectation of higher prices for change valuations is to be expected, even when the different effect of allowing for design is taken into account.

Valuing changes and provisional sum expenditure

General use of the contract sum analysis

The necessity of structuring the contract sum analysis pre-contractually to give the data to operate several provisions of the contract is developed further in Chapter 6, where the principles and problems involved are discussed. It must, however, now be taken as it is in terms of the distribution of amounts and what it is said that they cover. This is subject to the rider that 'everything is in somewhere' in an analysis which adds up to the contract sum, although both post-contract interpretation of wording to decide where and subsidiary analysis of amounts are in order.

In commenting on the various JCT clauses and also in Chapter 10, it is assumed that the analysis is perfectly straightforward and that the rules in each case can be applied without question. Several possible complications introduced in Chapter 6 over the analysis for any design and build contract need further treatment here.

All errors on the face of the analysis for a design and build contract, like other drawings and specification contracts, should have been cleared up pre-contractually, and if so the analysis in this respect is in order. If an error on the face is not noticed until the contract is under way, the parties should agree how to deal with it, perhaps in a similar way, while remembering that any tidying up of figures must not change the contract sum. All 'errors on the face' must be arithmetical: extensions, casts and so on. Unit prices and any quantities in the analysis are distinct and are discussed under the next heading.

Use of the analysis for valuing changes etc

The use of any analysis here centres around what JCT clause 12.5 terms 'the values of work ... in the Contract Sum Analysis', from which the values of changes may be deduced. The contract values may be presented as one or more of the following, discussed in Chapter 6:

(a) Amounts for sections on the lines of JCT Practice Note CD/1B, that is agglomerations of all work of one type.
(b) Amounts for sub-sections of (a), even down to individual items of work, but without quantities.
(c) Priced quantities, representing a refinement of (b).

It has already been indicated that the last of these presents problems in use, however appealing the array of detail may appear. This is because the quantities are presented by the contractor, and not prepared independently by a quantity surveyor acting between the parties, as in other arrangements. They are thus not open to any correction of errors resulting in an adjustment of the contract sum, because they are not 'guaranteed' independently. To check and agree the quantities before entering into the contract is a formidable task in time and effort, and so they are likely to be taken into the contract as uncertain. They therefore import corresponding uncertainty over how to use them in valuing changes, as is argued under the next heading.

To prepare firm quantities, it is always necessary to have firm design information – whatever may be said or even thought otherwise in the industry! If the design, as distinct from some details of specification, is incomplete at the date of tender, firm quantities in any conventional form are not possible. Only global fixed amounts are likely to appear in the analysis.

The analysis with quantities
As the use of either of the other methods tends towards some form of post-contract quantification, the quantities method (c) of analysis is considered first to bring out the basic points. A simple example of one section of work illustrates these points:

Item	*Quantity*	*Price £*	*Value £*
100 mm wall	20 m^2	30.00	600.00
200 mm wall	10 m^2	60.00	600.00
		Section amount	£1,200.00

This contains two unarguably fixed elements, so far as the content of the contract sum is concerned: the items as descriptions of the content of the section, and the section amount. The values may be agreed to be fixed as sub-section amounts, but see below. In an ordinary with-quantities contract the prices would be fixed, but here they are problematic because the quantities are as well: if one is 'wrong' so is the other, because the value is fixed. For example, if it were found post-contractually that the first quantity of 20 m^2 should read 40 m^2, a correction would mean that the price would have to become £15.00 to maintain a sub-section value of £600.00. As an error of this magnitude would distort the pricing (how should a 150 mm wall be priced?), any correction considered might be spread over both sub-section values, either by amending the prices to £20.00 and £40.00, or by arriving at a section amount of £1,800.00 less one-third. The mention of a correction anticipates the following discussion.

If a change occurs affecting a section which is then found to have a quantity error, there are several possible ways of using the data. In the foregoing illustration, a quantity of 20 m^2 instead of 40 m^2 may be termed a 'major error', while if it is 20 m^2 instead of 18 or 22 m^2 this may be termed 'a minor error'. Using this arbitrary distinction, there are several main possibilities which overlap and may be combined to some extent, or used separately in different sections of the analysis:

(a) Ignore the quantity error, whether it is major or minor, and measure the actual quantity variation and price this at the unit price given. This leaves the effects of the original distortion with the contractor (for better or for worse), obviates the need to check any original quantities and assumes that the unit prices are correct. There is an obvious question-mark over what should happen if a measured omission is larger than the contract quantity. It also means that there is no point in having the quantities included at all: a simple schedule of rates (the 'item' and 'price' column of the example) is all that is needed. This is the normal drawings and specification contract approach.
(b) Check all the quantities and make internal adjustments as described above

while maintaining the same section amounts, and then measure and price as in (a). Both the checking and (possibly) the adjustments are laborious. They could be eased by making the checking close but approximate, or by making adjustments to correct only major errors. This gives a more balanced basis for valuing changes, at least in the purist's eyes, by better proportioning from the original sections to the revised amounts. It also means that there is still only a limited value in including quantities, if these are open to such cavalier adjustment. This method is little different in principle from having only section amounts and analysing these into quantities and prices, when and as necessary, as mentioned below. It simply means that one party has given his ideas in advance, rather than both agreeing later, which may then save some work. The employer may well prefer such later composition, in that he can influence the detailed presentation more, or his consultant quantity surveyor can.

(c) Ignore the quantity error if either it or the quantity variation is minor. Proceed then as (a), but if the error and the variation are *both* major, deal with the quantities and proceed as in (b). This is obviously a compromise which may be 'fairer' to the parties, but which begs the question of what is 'major' and what is 'minor'.

Each of these options has been stated rather starkly, as though the contractor's quantities are bound to be seriously flawed. This is by no means the most likely situation. The fundamental point that they demonstrate is the limited contractual value of the quantities which are not accepted by both parties as accurate, subject to the right to adjust the contract sum to take care of any errors found. In each case the fall-back position is really a schedule of rates or a plain lump sum open to analysis. It may be doubted whether quantities are really worth while as something to fall back on, when the way back may become like the road from Moscow. If they are required, the employer's requirements should give rules on how to treat inaccuracies, preferably with flexibility.

Not only are quantities of limited value for straightforward use but, in a dubious case, there is more scope to try 'variation (or in this instance "change") spotting'. This is the practice when pricing traditional bills of quantities of putting high unit prices against items for which additions are expected (either because of errors in the bills or expected variations) and low prices for omissions. There it can be done only in pricing, whereas here the quantities are open to manipulation, with the resulting unit price swing not looking so extreme. The whole practice is a risky one for any contractor to use, but then contracting is inherently risky.

The analysis without quantities

From reading between the lines this, it must be underscored, appears to be what the JCT expects in its practice note and in its clause 12. This interpretation is supported by the discussion of the inadequacies of contractual quantities under the previous heading.

The JCT list of sections in the practice note is considered in Chapter 6. A particular analysis might consist of a dozen to fifty sections, each with a single price. This is alternative (a) set out before discussing the analysis with quantities. Alternative (b) is the same analysis, but with some or all of its sections divided into sub-sections, each with a single price. The extent of division may go down as far as it does when quantities are used, but with only lump sums given, not even an indication of quantity. Thus the example under the analysis with quantities would be simplified to:

Item	*Value £*
100 mm wall	600.00
200 mm wall	600.00
Section amount	£1,200.00

Alternative (a) would give only the section title 'Walls' and the amount of £1,200.00. Any analysis is available for such other purposes as interim payment and formula fluctuations calculations, and these too may influence the level of detail.

For valuing changes, the JCT list may be considered too coarse-grained in cases of complex changes, so that alternative (b) may be introduced where these are likely. The more detail there is and the more prices there are, the more is it possible to adduce some criticisms of the types already used against quantities. However, these are not so cogent because the fundamental problem is quantities on the face of the document which can be shown categorically to be incorrect. If prices are given with no means of *proving* them wrong, they must be maintained. In general of course *any* analysis can be criticised in some way. If only alternative (a) is used, an instructed change which increased the area of 100 mm wall by half would lead to a subsidiary analysis to produce something like the example given previously. To achieve this, two sets of variables are in play: the relative areas of each wall and the relative unit prices. It is by no means clear-cut that one price will be *exactly* twice the other, as shown. If say ten items are included in the section, but only two changes in quantity, extensive quantitative analysis of all ten is needed before money values can be allocated. Alternative (b) helps here, given prior subsidiary pricing, although the 'error' problem is not eliminated.

The alternative basis for changes in a design and build contract is to use a schedule of rates, as in any other drawings and specification contract. This is not available under the JCT contract, where clause 12 requires valuation of work to relate to 'values' in the contract sum analysis. Indeed, with a schedule of rates, an analysis is unnecessary – at least for present purposes.

JCT clause 12(3.6, 3.7 and 4.5) – Changes in the employer's requirements and provisional sums

Definitions

There is strictly only one definition, that of 'change' in clause 12.1, but it may be noted that clauses 12.3 and 12.4 both restrict provisional sums to any included in the employer's requirements, a point explained in Chapter 6.

The word 'changes' is used in clause 12.1 and elsewhere as an alternative to the full term 'Change in the Employer's Requirements'. This emphasises that the origin of changes is with the employer. Even if the contractor considers that something more than developing what, in his embryo design, is desirable, he cannot introduce it and secure an adjustment of price by himself. He must persuade the employer, if he will, to instruct a change. But it also emphasises that the employer cannot change the contractor's design directly and in detail, which would compromise the contractor's responsibility for his design. The employer can only change his own requirements, leaving it then to the contractor to modify his proposals in his own way in response. Special positions arise over discrepancies under clause 2.4 and over changes in statutory requirements under clause 6.3 which may lead either party to seek a change. Comments under 'The use of changes' earlier should be noted over what follows here.

Clause 12.1.1 defines one class of change accordingly: that affecting the finished, physical works. This is described generally as 'alteration or modification of the design, quality or quantity' on the contractor's part and made necessary by the employer's prior instruction. These terms range quite widely and overlap to some extent. If 'design' is affected, the contractor's consent is required under clause 12.2, discussed below. The definition is illustrated by the list given as 'including', but not restricted to the three categories in it. Inevitably, these again overlap. The first two are essentially changes in quantity and quality which could affect the design. The inference is that they are introduced *before* the originally intended work is performed: the third category of 'removal from site' allows for later changes. Once work and materials are removed, the possibility of substituted alternatives is available. Apart from express changes of these types, others listed in Table 7.2 may be deemed to have been instructed and so be eligible for valuation. The contractor should ensure that the employer is always aware of these to avoid queries in agreeing the final account. On the other hand, a substitution of materials etc under clause 8.1 does not constitute a change here, even though the employer's consent is required.

Two exceptions from the definition are given: work of rectification under clause 8.4 and removal of what is not in accordance with the contract, as a prelude to replacement. These are largely coextensive but under this contract may lead without a change to an 'alteration or modification' to satisfy the original employer's requirements. In any case, no difference in payment to the contractor results.

Clause 12.1.2 defines the other class of change, again to the employer's requirements, but not affecting the finished works. This is described as 'the imposition ... of any obligations or restrictions ... or the addition to or omission of such obligations or restrictions' and this case is limited to the four categories given, as these are not simply illustrative. These all affect the contractor's manner of working, if they cause him to depart from his own optimum method, and potentially overlap. They may as originally given affect his proposals, if he is asked for a method statement or programme as part of those proposals.

In this case the contractor's consent is not called for under clause 12.2, because design is not directly affected (except possibly that of temporary items, which are not part of 'the Works' themselves), but the employer's wider interests presumably are. However, clause 4.1.1, empowering the employer to issue instructions, singles out those within clause 12.1.2 as those with which the contractor need not comply if he makes reasonable objection. 'Reasonable' is not defined because of the wide range of possibilities, but signals the possibility of arbitration which occurs at once by clause 39.2, without supplying the arbitrator with any guidelines. The obvious avenue of objection would be on grounds of substantial disruption of working method, destroying the main structure of the contractor's programme. This has to be interpreted and moderated against the background that the contractor is to be paid for the effects of a change. The other avenue is that a change of this type might have a consequential effect on design, if the order of erection of new work or demolition of old were affected. It may also be that two changes taken together have implications that neither has alone.

Instructions

Clause 12.2 empowers the employer to instruct changes as defined in clause 12.1 if he so wishes and adds the traditional statement about no change vitiating the contract. Between them they modify the 'entire' basis of the contract (see Ch. 1). This power is restricted by the rights of objection mentioned below, but also by the general principle that the contractor has contracted to provide certain works and cannot be required to provide something altogether different. Thus he cannot be required, without such adjustment of the basic contract terms as he may be prepared to negotiate, to build a church instead of an abattoir. Similarly he is not obliged to build two abattoirs instead of one, although a somewhat larger abattoir is reasonable. There is obviously a 'grey area' in matters of this sort. The employer must reasonably be allowed to have only half an abattoir built, with the price adjustment governed (as in the other cases) by the rules later in clause 12.

Clause 12.2 specifically requires the contractor's consent, not unreasonably delayed or withheld, to any change affecting his design. This must mean consent *before* the employer may 'effect a Change', if it is to be done with the contractor's consent and if delay can be prejudicial. The contractor should therefore have a procedure in writing, although this is not stated, for giving his consents so that he is not at fault. Here 'design' is to be distinguished from 'quality or quantity' in the finished works, which are listed with it in clause 12.1.1, but over which the contractor may not withhold his consent to a change. The possibility of the categories overlapping is touched on under clause 12.1.1, so that the contractor may well be able to demonstrate 'an alteration or modification in the design', an unqualified expression, in most cases when quality or quantity change. It may be the intention behind the clause that only changes affecting the basic nature of the contractor's design solution and the timing and extent of disruption are in view. This is not said, and a few examples illustrate the gradation of possibilities actually requiring his consent:

(a) Substitution of a large span prestressed concrete frame for one in steelwork, affecting roof pitches, coverings and high level services.

(b) Higher insulation standards leading to smaller boiler capacity.
(c) Better natural daylighting, running counter to (b).
(d) Greater upper floor loadings and additional hoist openings in the floors.
(e) Substitution of floor coverings with different wear characteristics.
(f) Substitution of external rendering for facing bricks.
(g) Substitution of one type of facing brick for another.

In example (a), the contractor *might* be on good ground in withholding consent 'reasonably'. His position in (b) to (d) is far more questionable. In examples (e) to (g), he is not likely to be able to refuse.

Whether the contractor can hold back consent to an essentially aesthetic design change is doubtful, however drastic it may be – unless, say, he can fall back on contravention of planning permission. An arbitrator could be in an impossible position deciding over artistic matters. The strongest base for objection may be that of technical problems. The contractor may not have the right expertise in his existing design team, or the construction capability available to him. Such problems *can* be overcome; it is a question of whether it is reasonable to do so within the contractor's total scale of operations or within any reasonable enlargement of the present project's programme. If the contractor foresees not a problem of execution, but one of subsequent performance (such as floors needing extra maintenance, or external walls not weathering so well), he must qualify his consent to cover his own liabilities and to warn the employer.

Changes involving 'quality' without 'design' may involve materials of similar type and performance, especially if only cosmetic functions are being performed. 'Quantity' without 'design' changes may mean higher or lower walls or the addition or omission of a bay of construction in a repetitive building. Even in this latter instance, loadings on heating and electrical systems can be drawn in as design aspects. The clause is uncertain in meaning and intention, but 'unreasonably withheld' is just as doubtful, so that it must be hoped that the good sense of the parties prevails in the absence of contractual fetters!

The second class of change, that to obligations and restrictions defined in clause 12.1.2, is not mentioned in clause 12.2. It is the one type of instruction under clause 4.1.1, the general provision for the contractor to comply with employer's instructions, with which the contractor need not comply if he makes a reasonable objection. Here he is bound to comply if he does not object, which is rather different from the first class, where he is due to consent before there is anything with which to comply.

Clause 12.3 is not optional in operation and the employer must instruct over the expenditure of any provisional sums. These can arise only because they are included in the employer's requirements, so that he has set up the situation in the first place. Nevertheless it is highly desirable that the employer and contractor should agree at an early date when information must be in the contractor's hands in each case. Even so, it is the contractor's responsibility to ask for it in time, in relation to the extension of time and loss and expense provisions (see clauses 25.4.6 and 26.2.7). Enough lead time must be allowed for the employer to resolve whether uncertainties led initially to the inclusion of provisional sums. Whatever instructions the employer issues, and this is an area allowing numerous possibilities, he must leave it to the contractor to perform design development and to integrate the work into the rest, unless something completely free-standing is being introduced. Even here, the employer must ensure that he carries the contractor with him. There is no stipulation here about the contractor giving his consent to what the employer requires, as over changes, but it is best for smooth running and final settlement if a similar procedure is instituted over all provisional sum work, which is like the 'additions' part of changes in its effects.

Valuation

The rules for valuation in clause 12.5 follow a similar pattern to those in other JCT forms and are intended to give a similar effect. They do not differentiate between changes in the works and expenditure of provisional sums on the one hand and changes in obligations and restrictions on the other. Their obvious thrust is in respect of the former, and discussion of the latter is reserved until later in this section.

'Additional and substituted work' is dealt with in clause 12.5.1, which is less detailed than other versions. Such work may arise out of changes or provisional sum expenditure. Where it is 'of a similar character' to work in the contract, valuation is to 'be consistent' with the value in the contract sum analysis. How this is done will depend on the nature of the analysis already discussed in this chapter and Chapter 6.

If there are detailed quantities, the method for most work is ready to hand, subject to the problems aired earlier in this chapter. If not it may be necessary to analyse further one or more amounts within the analysis to give data for the valuation. It is normal practice to measure extra work and price it on the basis of unit prices contained in the contract and this is the basic method here. Even if some proportioning method is used, such as 'there is 50% more of this piece of work, therefore the extra price is 50% of the original', this comes in principle to the same thing. The details of the measured analysis will vary according to the content of the contract sum analysis and the rules embodied in it, and must be interpreted against the background of standard quantity surveying techniques, unless the rules provide otherwise.

If the extra work is 'of a similar character' to that in the analysis, related prices should be used. Thus pipes of the same material and size attract the same price, while those differing in size are priced in proportion. Those in other materials have their prices adjusted for supply and fixing differences. The clause also requires allowance to be made for different conditions of working and 'significant change in the quantity'. The former might embrace position in the works, timing in the programme or more or less continuity or repetition for men and machines. The latter might cover delivery costs or size of plant used. The last provision is for a fair valuation of work not of a similar character. This might apply to the only plasterwork introduced into the works and the valuation should allow a reasonable profit margin. All the earlier provisions are intended, by the methods of calculation given, to give similar margins to those in the contract.

'The omission of work' is provided for quite succinctly in clause 12.5.2. All the values are in the contract sum analysis, even if some work is needed to disentangle them, and these apply directly to omissions.

Clause 12.5.3 deals with what the JCT practice note groups together as 'preliminaries' in outlining the contract sum analysis (see Ch. 6). These elements of management, temporary roads and services, scaffolding and so on are by way of being overheads of a somewhat inflexible nature for the project as a whole, or for extensive parts of it. The changes valued under clauses 12.5.1 and 12.5.2 may not affect the provision of preliminaries one way or the other. But if they do, an allowance may be made in the valuation. This is likely to be far less than directly proportionate to the rest of the valuation, although no general rule can be formulated.

When all these approaches of valuing work from what is in the contract sum analysis and from first principles or market prices are reviewed, there often remains extra work which is not amenable to any of them. Clause 12.5.4 provides for fair valuation by daywork, that is, payment of actual costs incurred for labour, plant and materials with additions for overheads and profit, to be used in such cases. The provisions are as in other JCT forms and no special questions arise. It is therefore necessary to include a schedule of daywork terms in the contract sum analysis, and to ensure that records are kept of hours and quantities put into daywork, and that these are produced and signed on time to avoid disputes. The employer and contractor should agree in advance when daywork is to be used. Occasionally it may be agreed to keep daywork sheets 'for record purposes only' when the basis of payment has not been settled and the best approach is unclear. This should be on the basis of no commitment, but does tend to create an implication that daywork (which may be dearer) will be used. It is therefore not a practice to indulge in regularly.

The rules so far considered for valuing extra work are intended to be applied in the order given to establish which is most suitable, that is which is closest to the terms in the contract sum analysis. This order may be summarised as follows:

(a) Values in the contract sum analysis for 'identical' work, that is closely similar.
(b) Values proportionate to those in the contract sum analysis, for other similar work.
(c) Values as (a) and (b), adjusted for working conditions or quantity change.
(d) Fair valuation of different work, by assessment at values in step with the contract terms and at reasonable normal levels in the industry.
(e) Any consequential adjustment of preliminaries, due to (a) to (d).
(f) Fair valuation of work impracticable to value otherwise, by daywork at terms included in the contract.

These rules, and that for omissions, constitute the mainstream provisions which deal with most changes etc in the physical works. There are three further matters mentioned directly in clause 12.5.

Firstly, clause 12.5.5 deals with a possible consequential effect of a change or provisional sum instruction. The introduction or withdrawal of work may affect the conditions under which work which is not changed in itself is executed. If so, this work may be revalued using the rules given above to reflect the difference. Examples are a residual quantity that is less economical, or work carrying only part of the cost of facilities which were to have been shared with work now omitted. Others would be work made difficult of access or piecemeal by the introduction of new work. Each of these could be matched by work made cheaper by opposite causes. Often these cases balance and it is sensible to let this happen and make only any compensating adjustment.

Secondly, clause 12.5.6 is something of a long-stop and deals with valuation which is not covered by any of the foregoing methods, which depend essentially upon measurement and pricing or upon daywork. It also refers to 'fair valuation', but clearly different in nature. It is to be used in two situations: one is where there is 'not . . . additional or substituted work or the omission of work', and the other where 'the valuation of any work or liabilities' for part of a change does not lend itself to regular treatment. The former case is directly that of a change in an obligation or restriction under clause 12.1.2. Here the effect might be a major once-and-for-all reorganisation of site working or a continuing change in productivity. The corresponding valuations would probably be by, variously, an unusual lump sum calculation and an analysis spread widely across the sums for work performed after the change. The latter case could relate to the 'preliminaries' element in a change or to some other lump amount, such as special transport.

The unique aspect of the present contract arrangement is the inclusion of design. Changes in design are mentioned in clause 12.4 and may be valued by any of the means given that happen to be suitable. The design aspect of changes or provisional sum expenditure may bear no directly proportionate relation to the building work involved. There may be extensive redesign to accommodate a change of small value, or just the reverse. With design preceding construction, a change instructed well in advance may or may not lead to the original design work being omitted before performance. Design is therefore most likely to be covered by clause 12.5.5. The separation of the cost of design in the contract sum analysis is obviously a great help in valuing changes, in which case they could be held to fall under clause 12.5.1. As so often, the clause is very obliging.

Thirdly, the proviso following clause 12.5.6 requires that nowhere in clause 12.5 is allowance to be made for loss and expense reimbursed under other clauses. The main possibilities here are clause 26 over disturbance, clause 28 over determination and clause 34 over antiquities. The first allows loss and expense due to the effect of instructions, rather than the cost of implementing them, and so includes the effect of changes. The intention is that changes are to be valued under clause 12 as though they were introduced in an orderly way, while any disturbance is valued under clause 26 etc. All of this appears to amount to little more than a reminder not to pay twice for the same thing. The only practical difference is that the value of changes includes fully for overheads and profit and is subject to retention in interim payments, while loss and expense are otherwise. In the case of obligations and restrictions in particular, it may be almost impossible and pointless to attempt segregation.

Giving effect
Clause 12.6 requires the additions and deductions resulting from valuations to become adjustments to the contract sum and so to go into the final account. Because of clause 3, they are also to be taken into account in interim payments.

Market and tax fluctuations in costs

JCT clause 35(4.6) – Fluctuations

This and the remaining clauses considered in this chapter deal with their subject matter by almost identical wording to that in other JCT contracts, leading to identical results in principle. They are concerned with detailed financial technicalities and, apart from clauses 38 and 14 (taken in that order) and also the supplemental agreement, are not affected by the design and build concept. Most of them are particularly long and tedious, but are adequately covered both at length and tediously in, for instance, the present author's *Building Contracts: A Practical Guide*. Compulsive readers may therefore be relieved to note that the present treatment is by way of introduction and outline only.

The purpose of the present, commendably short clause is simply to establish, by reference to the appendix, which one out of the following three alternative clauses applies over fluctuations. It is not the intention that no clause at all shall apply, and so no entry in the appendix will automatically select clause 36, which is the minimum provision. If, despite this, it is desired to have no provision, it must be written into the contract and at least clause 35 must be physically deleted. To avoid doubt, clauses 36 to 38 should preferably be deleted as well.

'Fluctuations' is the standard JCT term for increases and decreases in costs to the contractor during progress of various inputs for producing the works, which are paid to or allowed by him. For simplicity these are referred to as though they are all increases – which is not far out! The three substantive clauses may be distinguished.

Clause 36
Calculation of fluctuations here is made by reference to actual expenditure incurred by the contractor or his sub-contractors and without any reference to the details of the contract sum analysis, even though this is mentioned in the clause. The usual method is by use of time sheets for labour and invoices for materials to establish hours and quantities throughout the works. The levels of payment are governed by the rules in the clause. In outline these permit amounts due to changes in statutorily imposed matters only, so that the method is often known as 'limited fluctuations, based on costs' or something similar. It is intended for use

with projects of short duration or performed in relatively stable market conditions, where only the actions of government defy prediction.

Clause 37
Calculation is again made by the methods used for clause 36. In this clause the rules permit amounts due to statutory changes again, but also those due to market forces such as wage negotiations and price changes. As a result the term 'full fluctuations, based on costs' is applicable. It is intended for use with projects for which clause 36 is not adequate. Both these clauses require a great deal of detail work in their implementation, while the result depends on whether or not the contractor is efficient in using resources. On the other hand, they do reflect actual fluctuations in costs. It is possible to use part only, say 'labour' or 'materials', of one of them or to use complementary parts of each. The next clause stands rigidly apart from them: its basis is different, it is an 'all-or-nothing' clause, and it cannot be used in any way alongside one of these two.

Clause 38
The method of calculation here, often termed 'formula fluctuations', is by calculating adjustments to the values of items in the contract sum analysis to reflect market trends in prices for building work, as shown by regularly published national index numbers interpreted in accordance with a set of formula rules of a technical nature. These rules are published separately from the contract by the JCT and are incorporated by reference when this clause applies. The clause then gives the contractual basis and procedure. This system is not related to the costs directly incurred by the contractor and so not to his efficiency or inefficiency. This may be an advantage or disadvantage for either party, as may be the link to nationally reported tender prices which are affected by factors other than costs of production. In principle the method adjusts *all* costs and allows profit, as the others professedly do not. It usually involves less painstaking calculation, but depends upon a carefully structured contract sum analysis and a strict timetable of interim payments. JCT Practice Note CD/1B includes a section upon its application.

Clause 36 is contained in its entirety within the longer clause 37, and so the latter is dealt with first.

JCT clause 37 (Appendix 4B) – Labour and materials cost and tax fluctuations

Clauses 37.1, 37.2 and 37.4 follow a common pattern:

(a) The contract sum has been calculated in the way given and is to be adjusted on the happening of events specified.
(b) The basis of the contract sum and contract sum analysis is the levels of particular costs at the base date, which is to be stated for this purpose in the appendix, desirably as about ten days before the actual receipt date.

(c) Changes in these levels result in net additions to or deductions from the contract sum.

The difference between the three clauses is that they each deal with distinct elements of cost. Clauses 37.4 to 37.8 are then addenda relating to the first three parts.

Wages, emoluments and expenses
Clause 37.1 deals with changes in 'wages and other emoluments and expenses . . . payable . . . to or in respect of workpeople' and these are defined so as to include incentive payments and holiday schemes. Related to these, consequential changes in the amounts of employer's and third party insurances and contributions, levies and taxes are payable (these are calculated on percentage or other proportionate bases and so vary in *amount* without any change in rate). In addition changes in transport charges and fares lead to adjustment. The definition of 'workpeople' is limited to those in the direct employment of the contractor, covered by wage-fixing bodies and working on or off-site, so that self-employed persons and sub-contractors are excluded (the latter only are covered by clause 37.4). A formula is included to allow reimbursement on a proportionate time basis at craftsmen's rate for employees not covered by wage-fixing bodies, such as supervisory staff. These employees are allowable only if and while they spend at least two whole days a week on site, so that staff in distant design offices are excluded. All of these adjustments (except transport and fares) are related to a contract basis of what was 'promulgated at the Base Date' by the appropriate body, even if not then payable, so that the contractor has to exercise a measure of foresight in tendering.

Contributions, levies and taxes
Clause 37.2 deals with changes in 'types and rates of contribution, levy and tax payable by . . . an employer' and also changes in refunds of these amounts. This casts the net widely over all statutory impositions in the field, except that training board amounts are specifically excluded. It allows by 'types' for the introduction or removal of a class of contribution etc bodily. These definitions are to be distinguished from those in clause 37.1, which relate to changes in amounts dependent upon changes in wages rates etc and not upon changes etc given here. 'Contracted-out' pension schemes are brought within the clause to rank at the state level, or in some cases above it. 'Workpeople' are the same group as before, with the limitation to direct employees and the extension to those not covered by wage-fixing bodies. The difference here is that the base line for adjustments is given by those 'types and rates . . . payable . . . at the Base Date', so that no foresight is to be exercised.

Materials, electricity and fuels
Clause 37.3 deals with changes in 'market prices' of 'materials, goods, electricity'

and, if and only if allowed in the employer's requirements, 'fuels'. These changes are essentially in invoiced prices which cover both suppliers' own amounts and statutory impositions (other than value added tax which comes under clause 14 and the supplemental agreement). They do not provide suppliers with a blank cheque or allow the contractor to recoup the cost of switching to an expensive source of supply. Only 'market price' changes, difficult though the concept may be, are eligible. Again, prices 'current at the Base Date' form the basis, so corresponding to clause 37.2.

Calculation of fluctuation amounts flows from these rules, which contain much subsidiary closely worded detail, subject to the provisions later in clause 37. Such elements as wage rates, fares on public transport and national insurance contributions are of course publicly known or available and do not need to be written into the contract as amounts. Transport charges and prices for materials, goods, electricity and fuels are not constant and so are to be covered as those at the base date in lists attached to the contractor's proposals. Only those particular materials etc listed can be subject to price adjustment and fuels fall into a special category here.

Sub-contractors

Clause 37.4 stands apart from the rest of clause 37 by dealing with sub-contractors, but requires provisions as in the rest of clause 37 to be included in sub-contracts. When this is done, the contractor becomes entitled to adjustments for work which he sub-lets, at the same level as the amounts due to be passed on to sub-contractors. If sub-contractors' tenders are later than the contractor's, there is the possibility that increases will have occurred in the intervening period. Unless sub-contractors are instructed to base their tenders on the earlier main contract date, they will include these levels in their tenders, as the contractor has not. The contractor will then receive inadequate reimbursement, unless he has increased his tender in anticipation. If Sub-Contract DOM/2 is used (see Ch. 4), the terms of the incorporated Sub-Contract DOM/1 automatically meet the requirements of clause 37.4 here, although the defined base date will need amending in the light of these comments. This clause still does not cover labour-only persons, for whom *no* reimbursement reaches the contractor.

Procedures, stipulations and definitions

Clauses 37.5 to 37.7 contain an assortment of procedures, stipulations and definitions affecting what has gone before. Significant points, other than the completely self-evident, at this level of treatment are:

(a) The contractor must give timely notice under clauses 37.5.1 and 37.5.2 of any event leading to an adjustment, as a condition precedent to any increase (but not decrease!). This is a carry-over from other JCT forms, and is in them presumably to allow the architect to take any possible avoiding action by way of redesign. It may give the employer here the opportunity to instruct a change to this end, but his scope is much more limited with design in the contractor's hands. There is no obligation on the contractor to amend his design on his own initiative, and indeed he will not recover the costs of so doing without any instruction.

(b) The employer and contractor are allowed by clause 37.5.3 to agree what particular adjustments are to be 'for all the purposes of this Contract'. This again is a carry-over from a different situation, and appears superfluous as they have to agree everything anyway. The force of the words quoted is, however, that any agreement reached and so endorsed becomes conclusive in any proceedings and cannot be reopened by an arbitrator or the courts.

(c) Amounts are to be allowed in interim payments, if necessary based on provisional figures. This is the consequence of clause 37.5.4.

(d) Earlier clauses have referred to the 'net amount' of any adjustment, that is net of profit. Clause 37.5.6 reinforces this, although the comments on clause 37.8 should also be noted.

(e) If the contractor overruns the completion date, fluctuations adjustments are to be frozen at the levels then obtaining, by virtue of clause 37.5.7. This allows amounts still to go on accruing at those levels. The provision is no way retrospective: it does not allow any reduction of amounts already accrued, on the grounds that progress before the overrun completion date was also late. This provision applies according to clause 37.5.8 only if the employer has operated the provisions of clause 25 by fixing a revised completion date in response to every proper notification of the contractor. This date is still open to review after practical (and therefore actual) completion, but the employer must be up to date or he will have to concede fluctuations. If the date is reviewed in the contractor's favour, he receives any extra amounts becoming due.

(f) Daywork is priced at current rates and so is excluded here by clause 37.6.1.

(g) No temporary items, that is those not finally incorporated into the works, may be included, except 'timber used as formwork' according to clause 37.3.1. Most other items will show as ineligible if put forward in the contractor's basic price list and should have been deleted. If not, they are still excluded by the effect of clause 2.2.

General percentage addition

Clause 37.8 requires 'the percentage stated' to be added to all of the contractor's fluctuations amounts. Clause 37.4 covers the addition of the same percentage to sub-contractor's amounts. No explanation is given, but this arrangement allows the contractor to make an allowance for elements of fluctuating costs not covered by the preceding clauses. These include other workpeople, materials not meriting separate coverage, plant and other temporary items, overheads or even profit. This is a blanket approach, made even more so by common percentage across all the separately calculated elements. It would be possible to use differing percentages, if it was considered that greater finesse could be achieved.

JCT clause 36 (Appendix 4A) – Contribution, levy and tax fluctuations

As already indicated, this clause is more restricted in what it allows and covers only statutory impositions. In respect of labour, it has as clause 36.1 those parts of clause 37.1 which give definitions embodied among the various parts of clause 37.2, subject to a number of drafting points. The effect is that of clause 37.2, except that the sometimes higher recovery for 'contracted-out' pension schemes is not included. In respect of materials, there is a quite distinct clause 36.2 worded to cover statutory matters and referring to 'types and rates' after the manner of clause 37.2 over labour. The intention therefore is to allow only such adjustments, which means analysing figures like materials prices to isolate the parts concerned. The concept is clear, the execution is difficult, especially as taxes etc may occur at various points in a long supply line.

JCT clause 38 (Appendix 4C) – Use of price adjustment formulae

Scope

Clause 38.1 requires the contract sum to be adjusted in accordance with the whole of clause 38 and the formula rules to which it refers. This requires use of the contract sum analysis, considered further in its detail under clause 38.2. However, the provision is strangely incomplete. It refers to adjusting the contract sum, but omits any reference to adjusting the values of changes and provisional sum expenditure. These are required by clause 12 to be 'consistent with the values of work' in the contract sum analysis and so to be calculated without adjustment for fluctuations. The position is the same in statement under clauses 36 and 37, but there calculation of fluctuations is based upon an analysis of the contractor's underlying costs of all his work, so that reimbursement in practice covers changes and other adjustments, such as the on-site elements of loss and expense amounts. Under the present clause, it is perfectly feasible and only rational to apply formula adjustments to most changes and provisional sum work, and it is suggested that this should be done as being the presumed intention of the clause. All the same, a judicial decision would be interesting. The formula method is unsuited to various other adjustments of the contract sum, and such elements as changes consisting of completely novel work and payment of loss and expense should be valued on a 'current value' basis.

Method

There are two quite distinctive elements in using formula adjustment under this contract. One is the contract sum analysis in clause 38.2. This analysis is covered in general terms in Chapter 6 and in its use for valuing changes under clause 12 in this chapter. Its structure and detail for that purpose may be unsuited to the present, because a section of work may combine in one sum values belonging to more than one work category. If so, the contractor is to provide 'any amplification ... necessary'. The earlier this is done the better; ideally it should be performed pre-contractually, when there is no commitment. It cannot be done for any section while its design is undecided, except on a provisional basis. The clause requires the contractor to 'include a statement of the allocation of the values' with each application for interim payment, but this should not be read as allowing amplification of the analysis to be provided progressively at the last moment in each case. There is a clear risk of argument if work categories are undergoing index changes at varying rates, as the value of work included in each payment is allocated to work categories and the appropriate indexation is applied. Prior allocation to categories eliminates this area of potential dispute.

Timing

The other distinctive element also relates to interim payments and is not mentioned in this clause. It is the use of a 'mid-point' rule for the valuation period covered by the payment. The bulletins are published monthly, but the precise index number for each work category is selected to lie in the middle of the valuation period and to apply to all work in the period. Provided work is performed uniformly and the index numbers change evenly, this is accurate. Even if these conditions do not obtain, the level of accuracy is usually acceptable if each valuation period is fairly short, while the relatively large number of periods mean that inaccuracies tend to balance out. Longer valuation periods mean less periods as well, and greater risk of inaccurate distribution on both scores. In all cases, more slowly changing index numbers help even out the results. Stage payments under clause 30.2A usually produce these longer periods, while period payments under clause 30.2B are usually at one-month intervals.

The parties should therefore carefully consider the balance of advantages and risks when deciding on which system of fluctuations calculation to embody in the contract. The 'accuracy' of calculations needs to be seen against the background philosophies of the fluctuations systems outlined under clause 35. In the case of the formula system used with stage payments, there is an already 'inaccurate' system on some arguments, but is it necessarily made more so by stage payments? It is possible to take some of the bumps out of the ride by dividing stage payments into smaller parts in turn with separate index numbers, but such an arrangement must be included in the employer's requirements or contractor's proposals. It is introducing more work into what is intended as a streamlined system. The JCT practice note goes the other way and suggests the option of aggregating the work categories for a stage in advance and with an eye on changes into one work group, which is the alternative method under the formula rules.

Subsidiary provisions

The rest of clause 38 is following other JCT versions closely. If the employer is a local authority, there may be a 'Non-adjustable Element', that is a percentage abatement of fluctuations adjustment for 'efficiency' (this even applies to decreases!). Adjustments taking no account of value added tax are to be made in all payments and are final, subject to quite limited correction rules. There are provisions about fluctuations on imported articles and what stop-gap action to take if publication of the monthly bulletins is delayed or ceases. The rules about freezing the level of adjustment if the contractor overruns his completion date occur here again, with suitable modification of detail.

Clause 38.4 is also a carry-over from the other forms. It gives the employer and the contractor the right to agree to modify 'methods and procedures' under the clause. As they are the parties to the contract, this gratuitous provision must be seen as encouraging. As with the corresponding provisions in clauses 36 and 37, there is the statement that amounts so calculated shall hold 'for all the purposes of this Contract', so precluding review in proceedings.

In view of the final nature of amounts for formula fluctuations included in interim payments (quite apart from the special finality of clause 38.4) and the fact that the contractor produces all the figures, the employer's right of challenge under clause 30.3.4 is important, as are the rights of the contractor under clause 30.3.5.

Fiscal matters

JCT clause 14(5.3) – Value added tax – supplemental provisions

Both the clause and the provisions at the end of the document repeat what is in other JCT documents, subject to changes in detailed terminology. In outline, value added tax is excluded from the contract sum by clause 14.2 and from all adjustments to that sum by other clauses. The supplemental provisions require the employer to pay to the contractor tax alongside interim and final payments, subject to rights of objection, challenge and special arbitration. There are procedures about tax statements, invoices etc.

The special element in this contract is the inclusion of design, which would be subject to tax if it were a separate service by the consultant. If it is included within a lump sum contract it is not subject to tax, unless the building work itself is so subject. This applies to the position when design is not shown separately in the contract sum analysis. If design work is shown and priced separately in the analysis, the Commissioners of Customs and Excise have confirmed that this does not in itself make it subject to tax. They have indicated that liability follows that of the works concerned. This means that, if necessary, the design sum must be split to enable tax to be calculated and charged on the relevant portion. (See also 'Contract sum analysis' in Ch. 6.)

JCT clause 31(5.4) – Finance (No. 2) Act 1975 – statutory tax deduction scheme

This clause is also a repeat from other JCT forms, because its subject is a repeat: the legislation to deal with the 'lump' of self-employed labour that is common in the construction industry. It needs no special comment in the present context. It applies in the contract only if the employer is himself 'a contractor' within the meaning of the legislation, and so responsible for collecting tax. He is 'a contractor' if he himself engages in 'construction operations', which include work as part of other work and even what he does on his own premises. If so, *the* contractor becomes a sub-contractor for this purpose. He may even become one part way through the present contract.

The clause is concerned entirely with these relationships and responsibilities over collecting tax or demonstrating possession of a tax certificate and all the other apparatus of operation. It establishes certain contractual relationships over what are legal responsibilities, even without the clause.

Chapter 10

Payments, completion and settlement

Interim payments

Completion, satisfaction and liabilities

Final payment and settlement

The matters taken in this chapter draw together all the threads in the final reckoning. They start with the mechanisms for paying the contractor during and soon after construction, on a provisional basis for work and materials already supplied, but not in advance of supply. They follow through the immediate physical completion of work and the subsequent liabilities over defects and their remedying. These stretch beyond the final account and payment, which top up what has been paid on account. In the closing days of the contract at final payment there still remains the possibility of arbitration over disputed matters, which may not be completed until after the close proper. Well beyond lies the possibility of legal action by the employer for a number of years over some classes of defects in design and work.

The ending of the contract proper may creep up on the parties, especially the employer, quietly in the absence in this type of contract of an architect's final certificate to evidence that all has been done and that all should now be paid. The JCT form marks the end by the agreed final account and statement acting as conclusive evidence on other matters also. Alertness is needed as to what is happening here, particularly as agreement may effectively be deemed to have occurred by passage of time in some circumstances. This 'end' is also the end of the employer's opportunity to recover any matters by way of contra-charges, so that he is advised to clear up any such matters ahead of schedule.

These are some of the more important threads woven into the discussion that follows, and they should be looked for at various points, while other matters of detail are not neglected. Those parts of the chapter which deal directly with the JCT contract relate mainly to a large part of Section 4 and all of Section 9 of that contract in its anticipated section headed revision, while a few pieces are likely to be separated out and appear in Section 1 (revised numbers are in brackets and see also comparison table in the Index of JCT clauses).

Interim payments

It is standard practice for building contracts covering all but the smallest works to provide for payments on account to be made during and perhaps after construction. This is a modification of the 'entire contract' principle and clearly provides cash flow for the contractor and so obviates him having to include for heavy financing charges in the contract sum. The amounts paid are strictly for what has been done or provided, and not by way of advanced financing. They thus accrue somewhat behind physical progress, how far depending on which method is used, but not always behind when the contractor himself has to pay out to his suppliers and sub-contractors.

It is a further standard practice to withhold a small percentage of these interim amounts until the works are completed and then, as a further modification of the entirety principle, retain a proportion of this 'retention' until defects discovered during a period of some months have been remedied. The money held gives some safeguard to the employer against the contractor failing to remedy defects and also against him failing to complete the works at all. It may not be adequate, especially against the second contingency (see Ch. 8), but is in addition to other remedies that may be available. These include determination provisions in the contract and proceedings related to it. The greatest upset for the employer is if the contractor becomes insolvent and probably unable to perform or pay for the employer being out of pocket. The contractor has his own concerns too if the employer becomes insolvent.

There are two broad methods of arranging interim payments, both of which are applicable to design and build contracts. The one most commonly used in other contracts also is payment at regular intervals, usually monthly, during progress and perhaps after as well. In this method it is the practice to value the work performed and the material delivered since the last valuation and pay on this basis, subject to deduction of retention and a permitted time interval before payment is due.

Valuation for periodic payments is usually performed by reference to the contract sum analysis which should therefore be suitably divided to throw up work in packets reasonably related to progress, or at least susceptible to allocation between several payments. The 'value' of a packet of work at the time of an interim payment may well be less than the amount included in the contract sum analysis, because the work is inadequate, perhaps until other work has been done to support or protect it. Indeed JCT clause 30 does not require interim valuations to be related to the contract sum analysis at all (as it does require changes to be related in clause 12), any more than the JCT with quantities versions require the contract bills to be used. Only when formula fluctuations are to be calculated under clause 38 in relation to interim payments and the analysis, does it become at the very least sensible to use the same basis throughout.

The other method is to pay predetermined amounts when the work reaches predetermined stages. These amounts include for further materials delivered and are usually again subject to a deduction of retention and a time interval. This method may mean stages which are entirely discrete and so follow their predecessors quite rigidly. But for a project having several buildings or other such parts, there may be stages within each part, so that parallel stages may well overlap but not coincide. Standard contracts are not worded to provide for such complications and some amplification is needed. This situation appears to be envisaged by the footnotes to JCT Appendix 2 in the words 'by stages (including by quantity of units and sub-units completed)' and would occur with housing estates for instance; but nothing is prescribed except how to fill in the schedule of stages!

One of the aims of the stage payment method is to give the contractor an incentive to push progress up to a stage and so release the amount concerned. It also tends to produce less frequent payments, as well as saving much of the basic labour of calculation associated with the other method during progress. Sets of overlapping stages running in parallel within a contract tend to nullify these aims, and may produce far more interim payments, to everyone's confusion. Some balance may be struck here by having regular interim payments, but including only values of stages that are complete at the dates concerned. This requires even more amplification of standard provisions.

Even a simple stage payment system as envisaged by standard contracts does not avoid all work. The values of stages still have to be calculated before the contract is formalised, which first means a close examination of the construction programme to agree the parcels of work and their sequence. The result should then be set alongside the contract sum analysis (again not obligatory) to produce the values. Some construction elements as given in the analysis may straddle the stage divisions and so need splitting. This is bound to be the case for preliminaries and design, which are by way of overheads to the rest. The elements may also include provisional sums and these need to be identified. They may be excluded from the stage values, so that actual expenditure (see next paragraph) can be included without leading to duplication. The JCT contract requires the stages to add up to the contract sum analysis and so to include any provisional sums, a point taken under JCT clause 30.2A.

The actual stages arranged will depend on the nature of the project: they should be clear-cut as to their occurrence, while allowing for incidental elements partly completed and for unfixed materials. It is also desirable not to extend too much the intervals between payments, irregular though these will be. A *maximum* of two months should be the aim for *most* stages.

In both methods it is desirable to allow amounts calculated during progress for adjustments to the contract sum, some of which may be deductions. The more important are changes, provisional sum expenditure, loss and expense amounts and fluctuations. In the nature of things, these amounts cannot be allowed with sufficient accuracy in advance in the stage payment approach.

If formula fluctuations apply it may be necessary to produce some subsidiary detail within the contract sum analysis, preferably before entering into the contract. It is also necessary to relate the valuation dates leading up to payments carefully to the mid-point dates for using the index numbers, which involves a greater problem of accuracy when the potentially longer stage payment periods apply. These points are considered under JCT clause 38 in Chapter 9.

The present chapter concentrates on JCT clause 30, which deals with all aspects of payment. The first part of the clause deals with interim payments, and so follows immediately, while the latter part deals with final payment and is taken towards the end of the chapter.

JCT clause 30(4.3 and parts 4.2 and 4.4) – Payments (Part 1: interim)

A comparison is given in Table 10.1 between the whole of the present clause and the JCT with quantities version. The rest of the present clause is taken later in this chapter.

Presentation and payment

Clause 30.1 sets out that clauses 30.1 to 30.4 govern all interim payments, but that within these one of the following applies:

(a) Clause 30.2A with alternative A in appendix 2, for stage payments.
(b) Clause 30.2B with alternative B in appendix 2, for periodic payments.

There are several common features to these alternatives. The general structure for an interim payment deduced from clauses 30.1, 30.2A and 30.2B is as in the JCT standard forms:

(a) Gross valuation to date:
  (i) amounts subject to retention;
  (ii) amounts not subject to retention, additions and deductions.
(b) Less:
  (i) retention on total of (a) (i);
  (ii) amounts paid in previous payments.

The end-product is termed 'the amount due', which equates with 'any monies due' the amounts from which the employer may deduct his entitlements by way of contra-charges, as allowed by the clauses listed in Table 6.2.

In each case the contractor is required by clause 30.3.2 to provide details as stated in the employer's requirements with the application for payment. This is different from the JCT standard forms under which the architect issues a certificate based, it must be assumed, either on his own findings, or upon a valuation prepared by the quantity surveyor, whatever preparatory detail may be provided by the contractor. Here, it is important for the employer to consider in advance what amount of detail he will reasonably require to enable him to check the application. This varies with the alternative applying and is looked at under each below, as is the timing of payments. In the nature of the contract, there is no reference at all to sub-contractors in this clause, just as there is no reference to domestic sub-contractors in JCT standard form clauses on payments.

Table 10.1 Comparison of clause 30 (payments) in JCT forms with design and with quantities

With design		*With quantities*	
Clause	*Subject*	*Clause*	*Subject*
30.1.1	Interim payments, obligation	30.1.1	Interim payments, obligation
30.1.2	Amount due	30.2	Amount due
30.2A*	Gross valuation, stage payments	30.2*	Gross valuation
30.2B*	Gross valuation, periodic payments		
Appendix 2, Alternative B	Off-site materials	30.3	Off-site materials
30.3.1/2	Contractor's applications	30.1.2	Interim valuations
		30.1.3	Issue of certificates
30.3.3	Employer's payment	30.1.1	Employer's payment
30.3.4	Employer's adjustments		
30.4.1	Retention deduction	30.4	Retention deduction
30.4.2/3	Retention rules	30.5	Retention rules
30.5.1	Final account to employer	30.6.1	Documents from contractor, final valuations from quantity surveyor
30.5.2	Adjustment of contract sum	30.6.2	Adjustment of contract sum and details
30.5.3	Details of last adjustment	30.7	Final payment of nominated sub-contractors
30.5.4	Final statement	30.8	Final certificate
30.5.5	Conclusiveness of foregoing account and statement		
30.5.6	Failure of contractor over account and statement		
30.5.7	Employer's account and statement		
30.5.8	Conclusiveness of last account and statement		
30.6	Balance due		
30.7	Notice by employer of deductions		
30.8.1	Effect of account and statement	30.9.1	Effect of certificate
30.8.2	Effect of proceedings	30.9.2	Effect of proceedings
30.8.3	Account and statement conditional	30.9.3	Certificate conditional
30.9	Effect of other payments	30.10	Effect of other certificates

*These clauses each contain the following sub-clauses:
1. Amounts subject to retention.
2. Amounts not subject to retention, additions.
3. Amounts not subject to retention, deductions.
In addition, clause 30.2B contains:
4. Special provision over certain materials.
Subsidiary elements in each clause are not detailed.

Under clause 30.3.3, the employer is to pay the amount due within 14 days of the issue of an application. This is obviously on condition that the application is not issued ahead of the proper date under the relevant alternative. It is also subject to the possibility, under clause 30.3.4, that the employer may disagree that the amount stated as due is contractually correct. This is distinct from deductions from the amount due, arising under clauses referred to above. If he disagrees 'on receipt of any Application . . . he shall forthwith' notify the contractor 'and shall pay at the same time' what 'he considers to be properly due'. This appears to abrogate the 14 days by requiring the employer to consider *and decide* 'on receipt' and pay (less) 'forthwith'. It is suggested that 'on receipt' is to be interpreted as 'having received', placing 'considers' at any time within the 14 days and leaving 'forthwith' thus contingent upon an uncertainty. It is to be hoped that any judicial interpretation would be similar. Even without any tremors of disagreement, the period of 14 days for checking and payment is far tighter than that in the JCT standard forms, where the employer has 14 days simply to pay on an architect's certificate. That certificate need not take into account work done less than 7 days before its issue and may have been longer still being compiled.

The employer does not have to agree any reduction with the contractor, or even warn him that it is proposed: he merely pays and explains. The contractor, however, has his right preserved in clause 30.3.5 to immediate arbitration under clause 39.2.1, where the same terms are used as in the clause: 'improperly withheld' and 'not in accordance', suggesting undue deduction and miscalculation respectively. The latter is not easy to foresee in the present contract, unless it is complete non-payment of an amount, when and only when clause 28.2.1.1 may be invoked after notice to determine against the employer, as the ultimate weapon other than proceedings. In practice there is really room for the parties agreeing, so far as they can, the amount for any application before it is formally made, rather than have these sudden death arrangements. This is the common process with JCT standard forms, whatever the letter appears to say.

The general legal principle over interim payments is that amounts are only provisional and inclusion of particular extra items is not an agreement eventually to pay them at all. The negative expression clause in 30.3.4 'is not in accordance' does not mean that the employer finally agrees the rest of the interim application, or indeed any others that he does not reduce. The position of the contractor over formula fluctuations based on disputed amounts, which may later be reinstated, is protected by the 'without prejudice' of clause 30.3.5. Nevertheless, the parties may both strengthen their position by making these points explicitly, variously with the first payment and after the first reduction.

The distinctive features of the alternatives are taken separately, dealing with the second one first.

Periodic payments

This is alternative B and it is taken first because it is the standard approach in most JCT forms and so by familiarity is more simple to introduce. Clause 30.3.1.2 governs when the contractor may (it is stated 'shall') apply for interim payments. Fixed intervals are laid down in appendix 2 and the usual period given is calendar monthly running to include the period during which the employer issues the statement of practical completion. The payment for this last period reduces the retention held. Thereafter the contractor 'may', or again 'shall', apply not more frequently than monthly as further amounts are due, which really means as they are agreed while being already due. In particular there is a payment when the rest of the retention is paid over, as defects liability is cleared on the later of the two dates given. In no case does the smallness of the amount due act as a bar to an interim payment. As between the last interim and the final payment, not even the one-month interval applies.

After its introduction providing the structure set out above, clause 30.2B giving the contents of payments falls into three main parts. Clause 30.2B.1 deals with routine items which are subject to retention. The largest element is usually 'work properly executed', a term which allows the employer to exclude defective work. This includes work in changes and in expending provisional sums, but excludes work consequent upon damage by the specified perils and paid for under the insurance provisions. Except in a major case, removal costs of defective work might be left undeducted, as ongoing work soon covers this sort of minor item.

In the present clause, there is the addition of 'including any design work carried out'. So given, this means that if needs be, design work is to be paid for ahead of the construction to which it relates. This is particularly relevant for the major element of design that usually precedes the first interim payment. If the value of the design has not been isolated in the contract sum analysis, some further analysis will be needed. This exemplifies the virtue of asking in the employer's requirements for supporting details accompanying applications to be based upon the contract sum analysis adequately divided (see clause 30.3.2 and earlier comments on using the analysis, which may be compared with those in Chs 6 and 9).

Materials delivered to site are usually quite significant, so that there are two provisions: that they have not been delivered too early, and that they are properly protected. Early delivery has to be balanced by the employer against possible savings in price due to early purchase, which reduce fluctuations amounts if clause 36 or 37 is being used.

Payment for materials not on site is entirely at the employer's discretion, exercised from time to time, unless special provisions are included in the employer's requirements or contractor's proposals, requiring materials of some types (perhaps where there is extensive prefabrication or restricted site storage) to be paid for in this way as an obligatory matter. If payments are made, the extra clauses in alternative A of appendix 2 are imported tortuously by alternative B of the appendix to introduce protection for the employer against an inadequate title to goods if the contractor becomes insolvent: a protection which is usually but not always adequate. Clause 30.2B.1.3 simply mentions the exercise of the employer's discretion referred to there. If the employer exercises his discretion entirely on the lines given, this is sufficient on most points, as appendix clauses 1.1 to 1.9 are all matters on which the contractor has to satisfy him, before the value of the materials may be included in an amount due for payment. These clauses, which are included in the body of the JCT standard conditions for use at the architect's discretion, cover the physical condition and legal status of the materials, the contractor's substantiation of these matters and the fact that his insurance covers the materials until they are delivered to site. In view of the optional nature of the provision and unless something special and mandatory has been written in elsewhere, the employer may impose any other conditions if he wishes before choosing to pay.

Clause 2 of the appendix is the equivalent of another part of the JCT standard forms. It provides that materials 'included in any Interim Payment', and so actually paid for, 'become the property of the Employer' (as clause 15 provides over materials on site). This in itself adds little to the legal position over ownership and is really affirmatory, so as to introduce the wording restricting the contractor over moving the materials. The contractor remains liable for loss or damage and for the cost of storage, handling and insurance. The employer should abate his payments at this stage to the extent of delivery costs, while paying nothing for the other costs. Insurance is mentioned in two places in the appendix. In clause 1.9 it is to protect both the employer and contractor and is against 'the Specified Perils' (see Ch. 8). In clause 2 it is simply 'insurance' and follows the allocation of responsibility for loss and damage to the contractor. The contractor is advised to insure more adequately against his liabilities: the specified perils are by no means the most obvious hazards to be met on the road, and not the only ones off it!

In the first main part of the clause, there is lastly the amount of formula fluctuations. This will be the net amount to allow for deductions, or in theory a net deduction. This type of fluctuation is the only one subject to retention, presumably, if not logically, included here because it is by way of adjustment of values in the contract sum analysis. It is also the one which requires quite close precision over the timing and content of valuations (because it gives final calculations in interim payments), and the amount which should therefore be safeguarded if the employer's reduction of a payment is challenged by the contractor.

The second main part, clause 30.2B.2, gives additions amounts not subject to retention. These are the assorted and usually minor amounts arising out of the defects and royalties provisions, the possibly much larger amounts for loss and expense in general and due to antiquities, for insurance payments following damage by the specified perils and for fluctuations based on costs. These need no special comments, although it may be noted that the fluctuations here are treated differently from those mentioned above.

The last main part, clause 30.2B.3, gives any deductions for defects not remedied and fluctuations based on costs as the only deductions within the amount due not subject to retention. Other deductions under the conditions are from the amount *otherwise due* as the interim payment (see list in Table 6.2).

Stage payments

This is alternative A and is given following alternative B, so that it may be compared with it more easily. It leads, as already noted, to payments at more irregular and usually greater intervals than does the even, periodic method.

These intervals are related by clause 30.3.1.1 to completion of the stages of work set out in appendix 2, with the last stage following at practical completion and covering all parts of the work and reduction of the retention. Thereafter there is only one more interim payment: when the defects question is cleared as in the other alternative, and leading to payment of the rest of the retention and of any balances that have accrued over the past months. Otherwise the contractor must wait until final payment to receive anything further. It is not clear why this difference from alternative B should exist: the use of stage payments is no magic wand that ensures that changes, loss and expense, fluctuations etc will be agreed by the day of practical completion. But it is there and the contractor should ensure that all possible amounts are then brought in with the last stage, even if on a conservatively provisional basis.

The first three main parts of clause 30.2A correspond to those in alternative B. The first and major part is again the routine items subject to retention in clause 30.2A.1. 'The cumulative value at the relevant stage' is that set out in appendix 2, alternative A, where progressively larger amounts will therefore be included. A note to the final stage indicates that its value must equal the contract sum, so that all amounts in the contract sum analysis must be used as they stand and none of them adjusted in themselves for changes etc. The next amount in the payment is the 'valuation of Changes or of instructions' about expending 'provisional sums'. Presumably it is the work arising out of the instructions which is to be valued, rather than the instructions themselves. In view of the note in the appendix about the final stage equalling the contract sum, the instructions must be read to balance the books as saying something like 'omit provisional sum of £*x* for *y* work and add work to satisfy employer's requirements as follows'. In practice, the intention is clear.

These amounts add up to the equivalent of work including design and changes under alternative B, and perhaps also of unfixed materials. Materials on site are not mentioned here, but may have been included in the various stages entered in the appendix by some shifting forward of value from what follows. Alternatively, the stages may exclude any allowance for materials; either way, the parties have to agree the point pre-contractually. There is no authority to include separate extra amounts, except in the special cases covered by clause 30.2A.4. The work subject to retention is completed by the inclusion of the amount of formula fluctuations again.

There is no reference here, as in the alternative clause, to including 'work properly executed', so that defective work is by inference only excluded. There are in principle three options open to the employer if defective work is in an otherwise properly completed stage. If the extent is small, he may choose to ignore it, if other work executed belongs to a later stage and its value offsets the cost of remedial work. Again, if it is small, he may regard 'the amount stated as due' as not being 'in accordance with this Contract', and so pay a reduced amount under clause 30.3.4. But if the amount of defective work is significant, he may notify the contractor that the stage is not yet complete and that payment is not due at all. The contractor retains his rights by clause 30.3.5 in the latter two cases.

Materials off site are not mentioned in the clause either, but are dealt with by alternative A in appendix 2. This contains the optional additional provisions which have been mentioned as available for the periodic payments system. When used with stage payments as their position indicates to be their primary purpose, these provisions apply only when the values of specific off-site materials have been included in the table of the appendix and this has been endorsed accordingly. The arrangement therefore has to be made pre-contractually to be automatically available, although it could be introduced later by a supplementary agreement executed in the same way as the contract itself. The wording suggests that the materials are to be entered as a separate stage or stages in the appendix, but this need not prevent it or them being invoked for payment in whatever sequence with the rest is appropriate.

The next two main parts in clauses 30.2A.2 and 30.2A.3 are identical with those in alternative B, covering addition and deduction items not subject to retention, and need no further comment.

Clause 30.2A.4 introduces an extra point, both by comparison with alternative B and in the light of what might be expected from the rest of alternative A. It applies only when formula fluctuations under clause 38 apply and only when the specialist installations mentioned, lifts, steelwork or catering equipment, are included in the works. These are all cases where the materials content is a high proportion of the whole and where the materials are likely to be delivered in large consignments. It is therefore provided that the value of these materials shall be included in the interim payments, so that the formula rules can be applied to these amounts. This is made subject to the proviso used in clause 30.2B about not being delivered prematurely and being properly protected. This is not included earlier in this clause, as there is nothing there about separately identifiable payments for materials to which it can relate. Retention is withheld in line with other provisions.

This is an understandable arrangement, but it does mean that the cumulative values used in stage payments must be constructed differently when clause 38 applies, from when it does not. Otherwise there will be a temporary double payment at some stage. This is because the stage payments entered in the contract are to accumulate up to the total of the contract sum, while the valuation in clause 30.2A.4 'shall *also* include the value of materials' specified (emphasis added). The cumulative value of the stage in which it is expected (and no better can be done) that the materials will be included should therefore be assessed as though no such materials will be on site, and the value when fixing is due to have occurred as though they have been delivered and instantly fixed. Deduction of previous payments, including therefore materials value, then brings the total back to the correct level.

Retention rules
Whichever alternative applies over interim payments, there are essentially the same rules for calculating and dealing with retention in clause 30.4. The calculation rules are in clause 30.4.1 and give the usual JCT principles:

(a) There is a retention percentage of 5% in the absence of any entry in the appendix. The footnote suggesting 3% on the larger contracts is advisory and the percentage must go into the appendix to be effective.
(b) The full percentage is deducted from the appropriate total before practical completion. The various elements under the two alternatives are reiterated, except that formula fluctuations are not mentioned this time, which is untidy. If they count as 'work' here (the presumed explanation), they should be defined as 'work' in the earlier clauses.
(c) Half the percentage is deducted from the total for work between practical completion and clearing of making good defects. The same comment applies over formula fluctuations.

These rules leave some items in interim payments to be paid without deduction of retention, as earlier clauses require. They also mean that reduction and then elimination of retention depend respectively upon the employer's statement over practical completion and upon his notice of completion of making good under clause 16, assuming single step completion. If there is partial possession under clause 17, or if a sectional completion supplement is used, there will be two or more series of reduction and elimination of retention running, and usually overlapping.

If the employer exercises his right under clause 30.3.4 to pay a reduced interim amount, he has to 'issue ... a notice with reasons', but not necessarily with figures. In a simple case this will leave it obvious to the contractor how much retention is being held. In a complex case when the employer has disagreed over more than one item in an application this will not be so, if there are items subject and not subject to retention, and if the employer's 'reasons' do not include calculations. This uncertainty could become relevant if the employer becomes insolvent.

The employer's possible insolvency is guarded against by clause 30.4.2, which imparts a trust status to the retention in the contractor's favour. The employer is not obliged to invest the money and, if he does, he may retain the interest. The danger on insolvency is that of being unable to identify such trust monies, quite apart from the uncertainty just mentioned. The contractor may therefore require the employer to open a designated bank account for the retention, although the employer then has the benefit of any interest. This separate account provision is deleted if the employer is a local authority, presumably and dubiously on the basis that insolvency will not occur.

Despite all this, clause 30.4.3 emphasises the employer's right to make deductions from payments to the contractor, including from retention released in such an amount. The point of holding retention would otherwise be nullified. This is only when there is 'any right under this Contract', so that the employer cannot exercise any such right of set-off without contractual authority.

Completion, satisfaction and liabilities

Extended discussion of the contractor's responsibilities and potential liabilities over design etc, more particularly in the JCT context, occurs in Chapter 5 with relevant JCT clauses. Clauses dealing with construction and quality are considered in Chapter 7. This present section ends with the related clauses about practical completion and the making good of defects. The key issues in all of these areas may first be brought together here in summary and then their implications for residual liability beyond the ending of the contract noted.

Defects upon and after practical completion

During the progress of the works, the contractor is responsible for several salient matters:

(a) Completing the design and specification in accordance with his proposals, which are based upon the employer's requirements.
(b) Constructing the works in accordance with the design and specification and any changes introduced on the employer's instructions.
(c) Putting right any deficiencies, deviations and defects which he may notice, or which the employer may draw to his attention.

During the same period the employer should, in his own interests, be checking what is going on. He is not obliged to find *all* defects etc during progress, any more than he should positively approve what is being done, as has been said *ad nauseam* in earlier chapters. Much depends on the nature of the work:

(a) General layout and aesthetics: the employer must be held to approve these, as soon as he has examined relevant drawings etc.
(b) Functional facilities, such as heating, lighting and cooking equipment, and finishes, doors and windows: a similar position to (a), but the employer is not held to have approved technical functioning, only that the facilities seen *appear* suitable to do what he requires of them.
(c) Detailed design and specification of all work, including that in (b): the employer is not held to have approved this, unless he stipulated it precisely in his requirements.
(d) Work as carried out: the employer is under no obligation to inspect or test work during progress, although he may do so if he wishes and often will.

It is thus the ideal that the contractor should hand over to the employer a project complete and faultless, at what is usually termed 'practical completion', although the employer need not have contributed to this by his own inspection. The very term practical completion leaves room for possible inadequacies, as discussed under JCT clause 16 below. Contracts usually provide a 'snagging' procedure and period, as the same clause illustrates. This is normally as far as they go, but they do not supplant the wider legal position in so doing.

The defects liability period, as it is usually called, is a period within which the employer may find faults, which the contractor must then return and himself put right within a reasonable time. This will vary with the nature of the fault. The arrangement is superimposed on the longer period at law, the 'limitation period' (discussed below), during which the contractor is liable for the employer's financial loss, both in remedial work and consequences. This period is likely to run for several years beyond the formal end of the contract. At this point, signified in the JCT contract by the acceptance of the final account and final statement, liabilities do not change, unless there are aspects of the work over which the employer's 'reasonable satisfaction' is the criterion over quality and standards (see JCT clause 30.8.1.1 for details and for a warning about using the expression at all). On these aspects the employer is accepting that he is satisfied, so that the contractor's liability is extinguished over them. They fall otherwise into the same category as the rest of the matters discussed under this heading.

In legal principle, there is no distinction between defects of materials and workmanship in any building contract and those of design which are peculiar to design and build, so far as the contractor is concerned. This is one of the strengths of the arrangement: one contract, one package of responsibility. However, the practical distinctions made under this heading are valid, if elastic. When practical completion has taken place and the employer is using a building, he has the opportunity on performance matters to test whether it meets his requirements, as distinct from merely appearing to meet them: the position before occupation. He may, for instance, discover whether the warm-up period for the building and the temperature differential attainable is as he explicitly required, whether structural stability and thermal movement are within the proper limits or whether flooring meets maintenance criteria. If they are not, these are patent defects, those 'open to view', which should be listed in the schedule of defects and remedied before the notice of completion of making good defects is issued and, incidentally, the balance of retention is paid over.

If the employer does not deal with patent defects in this way, then by issuing the schedule of defects and the notice of making good, which declare that they are all listed and cleared, he loses his right to have them remedied. It is common under other building contracts to give the defects liability period as six months, although this may be given as twelve months specifically to allow for full operation of heating, ventilating and lighting systems. In such contracts, only materials and workmanship are under scrutiny, unless there is specialist subcontract design: the main design is the subject of a separate contract or contracts with consultants, with liability running separately. In design and build contracts, all liability is in one. It is therefore important that the employer should obtain an adequate opportunity to test his project for patent defects of an operational nature. For seasonally operating systems, this means twelve months. Somewhat less might be possible, according to the time of year that practical completion is to occur, and what the risk of late completion might be. It is strongly recommended therefore that the employer reviews his project and time-scale carefully and, if needs be, errs towards safety by stipulating the full period. He may be debarred later from asserting that an inadequate system was a latent defect, simply because he did not have the seasonal conditions to test it.

This is essentially a question of patent *design* adequacy over which the employer seeks protection. Latent defects, those concealed (not necessarily deliberately) and appearing later, are a distinct issue looked at below, whether they affect work, materials or design. As such, they are not peculiar to design and build contracts, but are nevertheless important.

Defects and damage after settlement of the contract

There may well be an intermediate time between clearance of defects and the ending of the contract with the final account and statement. Comments here apply as much to this period as to that afterwards. The only distinction is that the employer has not yet expressed any of the 'reasonable satisfaction' referred to under the last heading. To any hopefully marginal extent that this applies, the contractor remains at his maximum level of risk in this period – an incentive to clear the final account!

The defects that remain as liabilities are those that are latent, that is existing but not apparent. Normal deterioration of a building as it ages, along with wear and tear, do not fall into this category but must be borne by the employer. Unduly rapid deterioration or wear and tear, due to inadequate design or lower specification than proposed, or to materials or workmanship below the standard specified, will be evidence of latent defects, if the faults were not apparent during the defects liability period. The continuing, residual liabilities of the contractor do not constitute a period of free maintenance available to the employer. It is also likely that only the more major defects are going to be dealt with here, for two reasons: minor defects are often more difficult to distinguish from the effects of age while, unless the contractor accepts responsibility, the effort and cost to the employer of securing redress may be a disincentive. On the other hand, the level of consequential loss may justify legal action, if all gentler methods fail.

Whether liability is in contract or tort

The first line of enquiry for the courts is always in contract, as the special relationship deliberately established between the parties, who are the immediate persons in view over defects. It is possible that the parties may have included terms in the contract which limit the liability of one or both of them against what would otherwise be the position at law (especially tort), by decided cases or even (if permissible) by statute. Examples of this are discussed over design under 'Design liability' in Chapter 5, and see also the *Greater Nottingham* case under this heading. It also arises over materials and workmanship if the 'reasonable satisfaction' criterion applies, as discussed above and under JCT clause 30.8.1.

Allusion has also been made in Chapter 5 to the alternative of an action in the tort of negligence, pursued instead of or in addition to an action in contract. This is based on a duty of reasonable care owed by one person to another independently of contract. It may be both wider and narrower in subject matter, as noted below, and also involve more persons. It is, however, more difficult to establish liability.

In particular, in *Murphy* v. *Brentwood DC* (1990) and *D & F Estates* v. *Church Commissioners* (1988) the case of *Anns* v. *Merton Council* (1978) was reversed by the House of Lords. This all but extinguished the previously growing possibility of recovering in tort for economic loss suffered by the plaintiff without physical injury or damage and also removed the possibility of recovering for damage to the works themselves. The high-water mark of the economic loss doctrine was reached in the House of Lords in *Junior Books Ltd* v. *The Veitchi Co. Ltd* (1982) when a nominated sub-contractor in a non-design and build contract lost in action over defective workmanship without a design element and had to reimburse the consequences, which were entirely by way of economic loss. The case has been fairly consistently 'distinguished' by the lower courts, a term suggesting the judiciary disagrees with the decision while being unable formally to reverse it.

In view of the remarks in *Murphy* v. *Brentwood*, it is clear that the tortious liability of a builder solely as builder is now limited to personal injury and damage caused to other property by a structure constructed or being constructed being defective. There is not a liability in tort on a builder for the repair of the defective structure itself, nor for purely economic loss, such as loss of rent, flowing from the defects and their consequences. It is possible, although untested, that an architect or other designer may have a wider liability, including one for purely economic loss, as being a person having a duty of care on whom the plaintiff relied, following *Hedley Byrne* v. *Heller* (1964). If so, this may give a contractor performing design and build a liability over design faults as distinct from construction faults (with all the inherent problems of segregation), whether he is working with an externally engaged designer or not.

In *Greater Nottingham Co-operative* v. *Cementation* (1988), the existence of a collateral warranty with a nominated sub-contractor was held to exclude an action in tort over negligent performance, even though the matter in dispute was damage to adjoining property in separate ownership. Here the specific terms of the warranty (or contract) overrode possibly wider rights which might have existed in tort.

Who is liable and to whom

Contractual liability exists only between the parties to the contract, here employer and contractor. They in turn may have liabilities subsisting in either direction between them and others, typically the employer and his consultants, and the contractor and his sub-contractors and suppliers, including designers. There may also be liabilities to tenants, funders and others introduced by collateral warranties, which are ignored for immediate discussion but taken up later in this chapter. These may lead to one party to the main contract having a liability to the other. However, tortious liability may lead to a person being joined as defendant in an action brought by someone with whom he does not have a contract. In particular a construction sub-contractor or a design consultant may become liable to the employer or others if there is a duty of care. This liability could therefore still exist even when the contractor does not. Cases bearing on this liability, with the contract parties still in existence and ownership were discussed in Chapter 5, especially *Independent Broadcasting Authority* over design with an associated responsibility for installation.

When liability starts to run

This question has been under review in tort for a number of years and has passed through several phases. The main options have been that it runs from:

(a) The date when the breach occurred which has led to the damage occurring. This might easily lead to the limitation period of six years expiring before the damage has occurred or is known.
(b) The later date when the damage was suffered. It might again remain undiscovered until the limitation period has expired.
(c) The date when the damage was discovered or, if earlier, reasonably could have been discovered. This protects the employer against the weaknesses of (a) and (b), but exposes the contractor (or even his beneficiaries) to claims potentially without limit. It also takes matters beyond what it is reasonable for insurers to cover, which is usually the practical point of protection for the employer.

Matters turned the corner with *Pirelli* v. *Oscar Faber* (1982). Its facts illustrate the problems addressed by the Latent Damage Act 1986. Here consultants for the design of a concrete chimney were defendants against their client's action in tort alleging negligent design, resulting in severe internal cracking high up in the chimney lining. These facts were not disputed, but liability was, critical dates being:

July 1969: chimney constructed.
April 1970: latest date for cracking to occur, according to expert opinion.
October 1972: earliest proven date when cracks could reasonably have been discovered.
November 1977: cracks discovered.

With a six-year limitation period (see next heading), which of the central dates applied was crucial? It was held by the House of Lords that the period ran from the date when the cracking occurred, which was not later than April 1970, so that the action was time-barred.

Several decisions in subordinate courts followed *Pirelli*, such as *London Congregational* v. *Harriss & Harriss* (1985) and *Ketteman* v. *Hansel* (1985). There had been mention in *Pirelli* of the possibility of 'buildings doomed from the start', that is so defective that the cause of action in tort must be clear enough to run from practical completion, but this extraordinary suggestion has since been treated with reserve as likely to exist only in hypothesis and has effectively been consigned to oblivion.

The present position in tort is governed by the Latent Damage Act 1986, although frequently the parties (or at least the employer) seek to provide a contractual remedy by means of a collateral warranty, as discussed hereafter.

When liability is extinguished

The right of action over defects in design or construction runs or accrues for a defined time from when it comes into being. In the case of actions for breach of contract (including negligent performance), the right runs from when the breach occurred, which is usually taken as from practical completion, even though the actual action or inaction was most likely earlier. It then runs for six years for an oral contract or a contract under hand (simply signed) and twelve years for a contract executed with greater formality as a deed. The JCT forms all provide for either of the latter two options. Clearly, the last of the three gives the higher degree of protection so far as the employer is concerned. An oral contract is as binding at law as one in writing but, not to stress the point too highly, has problems of establishing its subject matter and terms with certainty.

In actions for the tort of negligence, the right runs from when the damage or loss is suffered and the period is then six years. This means that the right of action is usually ended at a date which is later than in the case of a breach of contract under hand, so that an action in tort may become preferable. It is limited, however, by not covering damage to the works themselves or economic loss.

The problem in contract or tort is that the right of action may have accrued and have been exhausted, that is have become statute-barred, before the defect is discovered. This is how matters rest for contract, where the limitations of tort over causes of action do not apply. As a sort of quid pro quo over tort, the Latent Damage Act 1986 has provided a wider framework, while protecting potential defendants from indefinite liability. It gives three main periods:

(a) Six years running from the date when damage occurred, as distinct from when the breach occurred, this being effectively what already existed.
(b) Three years running from the date of actual discovery or possible discoverability and applying over and above (a), this perhaps giving a later date for liability to end and not necessarily overlapping with the six years.
(c) Fifteen years running from the date of breach of duty (equivalent in tort here to 'breach of contract'), this acting as a 'long-stop' cutting both of the other periods short if needs be.

Several of the numerous possibilities are illustrated in Table 10.2. While the judgment in *Pirelli* would have been different had the Act then been law, it would appear that a principle from it remains valid. This is that a plaintiff may choose an action in tort when this is permissible and more advantageous. Often it will be, but it may not be available if a collateral warranty (another contract) is in force. This is raised below.

Collateral warranties

While the Latent Damage Act has strengthened the period available for many actions over defects, *Murphy* and its related cases have reduced the scope for tort actions by clawing back recovery of economic loss from the arena of litigation. There has therefore been a surge in the use of collateral warranties to seek to regain the advantage for building clients (employers in contract terms), particularly after the contract has ended, and extend it to others as well. Many of the issues surrounding these warranties are far wider and of more general application than for design and build work alone, but there are elements which are particularly relevant for outline discussion here.

With the restriction of tort actions, employers have turned back to contract actions as the primary port of call. These are fine and do allow recovery of economic loss as consequential, but are subject to the fact that the statutory limitation period may be less and, most critically, that an employer has privity of contract with his *contractor* but not with sub-contractors. They are on the far side of the employer and, in design and build, include consultants to the contractor. Provided the contractor is still in legal existence there is usually a line of action through him to sub-contractors, but with the disappearance of the contractor the line is broken and the employer is without remedy.

Further, there has been a growing demand from others on the far side of the *employer* from the contractor to have a line of redress against the contractor and against his sub-contractors, particularly against consultants. These others are especially prominent in development work: on the one hand there may be funding institutions and mortgagees who are risking their investment, while on the other there may be tenants of the employer or there may be a purchaser or successive purchasers from him and any related tenants. Some of these potential interested parties may be able to proceed against the employer over defects, leaving him to proceed along the chain against the contractor and others but, again, the ending of the employer's or contractor's legal existence would break the chain. The broad pattern which may come about is indicated by Table 10.3 in which the parties to a design and build contract are centred and others are set on the appropriate 'wing' and also grouped in their own column when they are similar or associated.

Collateral warranties may be set up here to skirt round the employer or the contractor or both. It may be discerned that quite a complex network can result and, when there is enough demand, often does! In some instances the employer may, as an alternative, be able to assign his main contractual rights to these parties, but this is not usually done because of legal complexities and because it is restricted by the need to obtain the consent of the contractor, who is not obliged to extend his line of liability.

Table 10.2 Duration and expiry of liability for latent defects

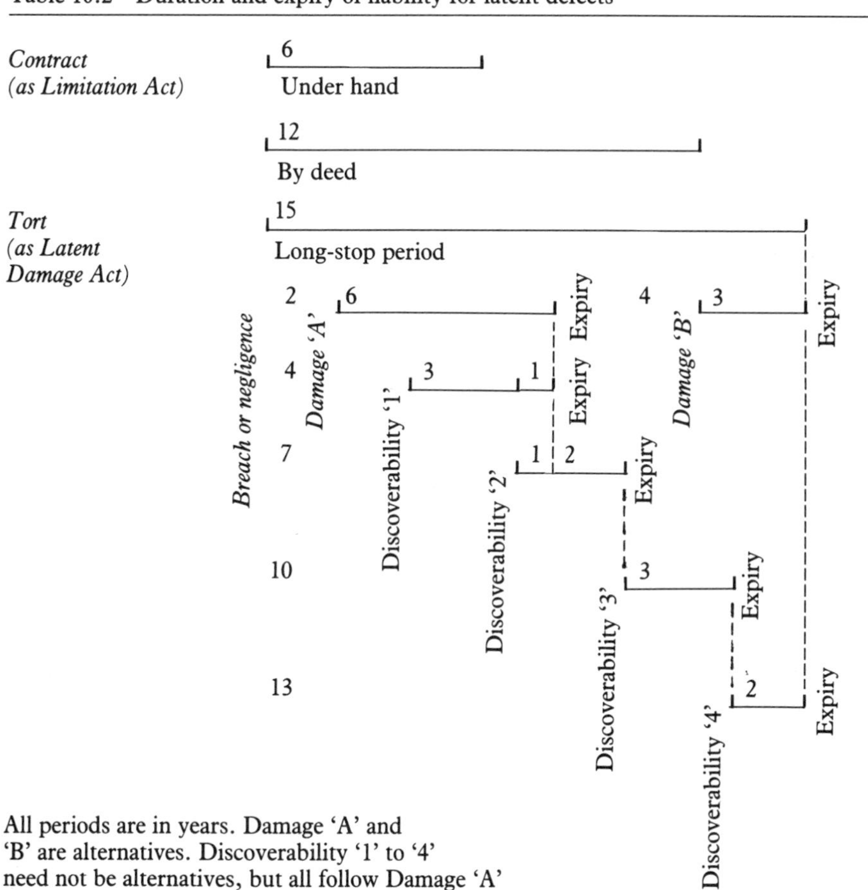

Table 10.3 Relationships in collateral warranties

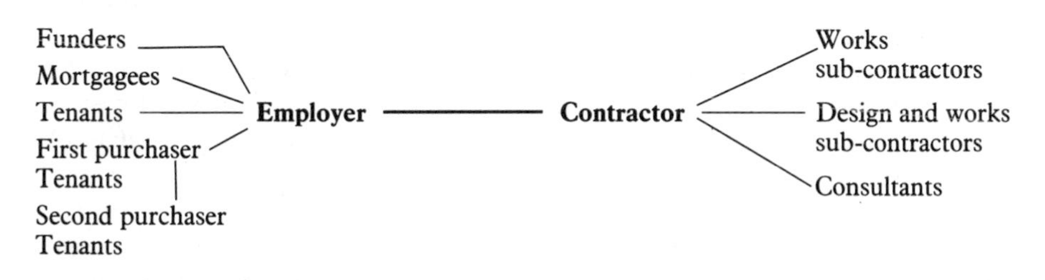

These warranties operate, as their name indicates, as agreements alongside the main contract to give a direct line of redress from one party to the other. It is possible for them to be set up with several parties to whom a joint warranty is given, although this is not too popular because the several parties may be in conflict when it comes to a settlement. They are very much one-way arrangements, in that there are no reciprocal responsibilities, and usually refer only to defects matters. To be enforceable they therefore need to embody at least a nominal consideration, payable to the warrantor, or to be executed as a deed.

Warranties collateral to another contract should properly be executed at the same time as that contract (which in the context of the building work may be a sub-contract) or be executed in advance and come into force when that contract does. This is because the warranty extends the responsibilities of the party beyond what is in the contract. Usually, warranties affecting those financing the project are ready to be cleared well ahead of entering into the main contract. However, it is not uncommon for an employer to seek warranties at any time after entering into the contract to satisfy some others of those on his wing in Table 10.3 and who may have been unknown earlier. There is no legal obligation on the contractor or his sub-contractors or consultants to do this and the matter is essentially one of commercial bargaining power (such as over future work). The contractor or other person should consider carefully whether the additional obligations which might be assumed are excessively demanding in direct terms of fulfilment, and also in terms of extending his insurance cover, as noted below. This may arise because of the extra persons brought into play and also because of an extension of the scope or nature of the work, perhaps in physical specification or commonly by elevating the standard of care.

Common examples of significantly onerous conditions are:

(a) That the works are fit for their purpose. This may vary according to the person receiving the warranty and so be difficult to apply unless the purpose is known in advance.
(b) No deleterious materials have been used, often with a list appended. Numerous technical problems of definition and borderline cases can arise here.
(c) Proper professional indemnity insurance is carried, with adequate limits of amounts. This will usually extend to consultants and, where appropriate, to sub-contractors.
(d) The contractor will execute further warranties in favour of persons (such as purchasers and tenants) yet to be named by the employer. This is by way of being a blank cheque and some limit to their interest or numbers should be included.

Such requirements as these should be cleared with the contractor's or sub-contractors' insurers before they are accepted, otherwise they will be left solely at risk. That of fitness for purpose is most unlikely to be accepted by an insurer. There are some fairly well accepted forms of warranty in existence, such as the BPF Form of Agreement (produced and issued with the agreement of several professional bodies), although none of these is specifically drafted to fit design and build situations. Even so, these should be examined with care and legal advice as to whether they in fact regulate in every respect the interests of each party in the actual circumstances of a project.

A warranty may not be an unmixed benefit for an employer. It may convert a tortious negligence claim into one in contract. This will remove matters from the scope of the Latent Damage Act, so reducing the period during which a right of action may accrue.

JCT clause 16(2.6 to 2.8) – Practical completion and defects liability period

This clause follows other JCT versions closely, with the omission of the usual last part, which simply restates the position over frost damage stated before. The employer takes the place of the architect at crucial points.

There is the usual vagueness over exactly what constitutes practical completion, which is commonly taken to be the juncture at which the works are believed to be complete and ready for occupation, although every detail has not been checked. In other JCT forms it occurs 'when in the opinion of the architect' it 'is achieved'. As he is absent, this subjectivism is also absent from clause 16.1. Instead, the works reach practical completion and 'the Employer shall give the Contractor a written statement to that effect'.

It is obvious that the contractor is bound to be the first to have an 'opinion' on this matter. He should advise the employer when it is drawing close, both for the implementation of this clause and to enable the employer to prepare to move in. It is, however, the employer who signifies practical completion and its date. He is not to delay or withhold the statement unreasonably, but there may be any arbitration, which may begin on all matters held over under clause 39.2, since there is now alleged practical completion of the works. If he is late with his statement, he should give the actual and earlier date. Most of its effects 'for all the purposes of the Contract' can adequately be retrospective, except any delay in the consequential interim payment. The 'purposes' that it affects are:

(a) Beginning of defects liability period and change in liability for frost damage under this clause.
(b) Ending of insurance of the works by the contractor, if clause 22 A applies, so that the employer should insure.
(c) End of liquidated damages accruing under clause 24.2.1, although those accrued may still be recovered.
(d) End of interim payments with the payment now due under clause 30.3.1.1 (stage payments), if it applies, except for one after making good defects.
(e) End with the next payment of regular interim payments under clause 30.3.1.2 (periodic payments), if it applies, with subsequent payments as amounts arise.

(f) Halving of retention under clause 30.4.1.3.
(g) Beginning of period for submission of final account and final statement under clause 30.5.1.
(h) Opening of any arbitration held back under clause 39.2.

It is also the case that the employer can no longer issue instructions over many matters, such as the performance of further new work, unless the contractor wishes to take them on appropriate terms.

Clauses 16.2 to 16.4 give the employer the power which elsewhere resides with the architect to instruct the contractor over remedying defects 'which shall appear within the Defects Liability Period', the length of which is to be given in the appendix, or otherwise in default of an entry, is six months. Most effects are as in other JCT forms:

(a) 'Which shall appear' embraces defects coming to light during the period, even if they were actually present before completion. In addition the employer may have notified minor defects to the contractor before completion, but have issued his statement on condition that they too would be remedied.
(b) The employer may issue *one* formal 'Schedule of Defects', so titled and constituting an instruction, and must do so not later than 14 days after the end of the period. If he is so minded, or maybe mindless, he may even issue it on the first day of the period. When issued, it ends his right to instruct over defects. He may, however, issue individual earlier instructions 'whenever he considers it necessary', provided he gives them some other title. These are useful over urgent matters.
(c) Any of these instructions is for 'any defects' etc to be 'made good' by the contractor. They therefore specify what is wrong, but not how the contractor is to put them right, otherwise the employer may find himself having to pay for the work. It is for the contractor to decide how to meet his contractual obligations, as much as during progress. This is particularly important in this type of contract, where there are differences mentioned below, as well as similarities.
(d) The contractor is to comply with these instructions 'within a reasonable time', both as regards starting and finishing. He is not an in-house maintenance man, unless something (very) special is written into the employer's requirements. If redesign is needed, this may well extend what is reasonable.
(e) The contractor does all work 'entirely at his own cost'. This is subject to the proviso that the employer may instruct that particular defects are not to be made good, when 'an appropriate deduction' is to be made when settling. This could occur because the employer is prepared simply to accept a lesser standard of provision, or because remedial work would cause too much upheaval in the works which he is now occupying. The contractor has no option about this and cannot insist that he should return to perform the work and avoid a deduction: the impact on both parties is to be allowed for in the

appropriateness of the deduction, but with the contractor at a disadvantage, as he has introduced the defects. Reasonably for defects of minor scale in relation to their deduction value, the contractor should be allowed to return. If the employer does choose to take a deduction, he should take account of the possibility of later problems with the work, which he will be held to have accepted by his action. The contractor may also be liable for consequential damage flowing from defects (cleared or otherwise), although the employer may need to bring an action for breach outside the scope of this clause.

(f) When *all* defects are cleared the employer is to 'issue a notice to that effect' under clause 16.4. This releases the contractor from liability necessarily to return and remedy defects himself, but not from his residual financial liabilities over defects which stretch up to the issue of the final account and final statement and beyond.

While these are essentially common effects with other JCT forms, with some difference in emphasis, there remains the difference in the scope of what constitutes 'a defect, shrinkage or other fault'. Pursuit of the internal distinctions here or in any contract is fruitless. The important issue is the contractor's responsibility over design and specification. Both before and during construction, the employer has been in the somewhat passive position of waiting to see that any drawings etc provided by the contractor 'appear' to be proper developments of the contractor's proposals, which he has already signified in the third recital of the articles 'appear to satisfy' his requirements. In continuation of this theme, he now has two distinct areas in which he may be looking for defects:

(a) Simple questions of poor materials or workmanship: bricks may crumble or flake, ceilings fall down or pipes leak. These are in common with other types of building contract.
(b) Complex questions where materials and workmanship may or may not be in order in themselves, but where the response is otherwise unsatisfactory to the employer.

What the employer may require to be remedied in the latter group as defects within the meaning of the present clause is largely dependent on the way in which he and the contractor have handled matters beforehand.

In outline, he is likely to be precluded from any complaint over matters of layout, visual impact and general usability, for the reasons already given in this chapter. If they were in the contractor's proposals, these are aspects on which the employer is bound by appearance while, if they were not, he had the opportunity to object if his essential requirements were not being satisfied, in the scheme development and in the response over changes.

On the other hand, he is normally quite justified in raising technical faults arising out of design. Here he is likely to have a poor case only if he has previously entered into a technical assessment and given categorical approval in that way, as it is indicated that he should not. Once more though, his instruction in the schedule etc must be in the form 'bring this matter into conformity with my requirements' and not 'take up flooring *x* and substitute flooring *y*'.

JCT clause 17(2.9) – Partial possession by employer

This clause presents only one point for discussion where its use differs significantly from that under other JCT contracts. Its purpose is to allow for some part of the works being completed ahead of the rest and being taken over by the employer with the contractor's agreement, without this arrangement having been written into the contract originally. It can be applied to several 'relevant parts' which successively pass the employer. The main provisions are:

(a) A statement issued by the contractor identifying the extent of the part and giving the date of possession.
(b) A separate defects liability period for the part.
(c) A separate notice issued by the employer when making good of defects is complete.
(d) Transfer to the employer of responsibility for insurance of the part, when the contractor has been responsible.
(e) Reduction of the periodic sum for liquidated damages for the rest of the works in proportion to the value of the relevant part within the contract sum and of the contract sum.

The last provision thus ignores any shift in the proportions of the values due to changes etc and retains the relationship which presumably was embodied in the contract. On the same basis, it also ignores any discrepancy between the contract values and the usefulness values to the employer of the several parts. This is quite regular JCT practice, unless any special values are included in the contract. This is unreasonably difficult to do with partial possession, as this is unforeseen and so unheralded. It is properly suited to sectional completion under the next heading.

Sectional completion supplement

This document is published for use with other JCT contracts to provide clauses giving a similar effect to partial possession, in those cases in which the employer requires phased completion to be written into the contract, naming sections and dates. It consists principally of minor drafting amendments to a number of clauses and so is not suitable for direct use with the present contract. The JCT practice note recommends that details of early possession be given in the employer's requirements and 'any necessary modifications made to the contract provisions'. The sectional completion supplement should be used as a guide for this, so that either a special version is produced or amendments are made directly in this contract. A supplement to the supplement is *not* recommended!

In no case can the contractor insist that the employer must take early possession of any *part*, although he is quite entitled to complete the *whole* works early and receive the benefits. In all cases, part or whole, these direct benefits are early reduction of retention and release from defects liability. No completion bonus, as a sort of inverted liquidated damages, is due.

Final payment and settlement

When any needed remedial work after practical completion has been completed, the stage is set for final payment and settlement, subject to any arbitration (considered at the end of this chapter) or legal proceedings. The precise sequence of events following completion varies slightly between contract forms. That in most JCT forms, with particular variations for the design and build forms, is shown in Table 10.4. It may be gleaned from preceding sections of this chapter but summarised as:

(a) Practical completion, with known balance and first half of retention paid in the next interim payment.
(b) Defects liability period, with any extra period needed to remedy defects, during which any further balances ascertained are paid, only if periodic interim payments apply.
(c) Notice that defects are made good, with second half of retention paid at once and thereafter any balances as in (b).

Thus retention is dependent on physical work and balances on financial calculations. If stage payments apply no balances are paid separately from retention, but are held over until the final payment. The final payment in the JCT pattern takes the place of the architect's final certificate under other forms, and has similar effects. It is contingent upon remedying of defects and agreement of the final account, perhaps by default, and its embodiment in the final statement. These final elements are considered under the rest of JCT clause 30 below.

Whatever the form of contract, there are broadly the same documents that the employer should obtain from the contractor to enable him to check the final account. A typical list is:

(a) Instructions and other authority for including items (see Table 7.2).
(b) Subsidiary analysis of items in the contract sum analysis, perhaps with quantities and unit prices (but see discussion in Ch. 9).
(c) Synthesis of values based on (b) or on 'fair valuation'.
(d) Sub-contractors' and suppliers' accounts in support of specialist work in (c), although the contractor may be justified in contending that it is commercially unreasonable to require some of these essentially domestic and private documents.
(e) Daywork sheets, authenticated and priced.
(f) Loss and expense data, if not already supplied.
(g) Fluctuations calculations.
(h) Miscellaneous details relating to adjustments for fees, insurances etc.

Table 10.4 Timetable for settlement of final account under clause 30 of JCT with-design form

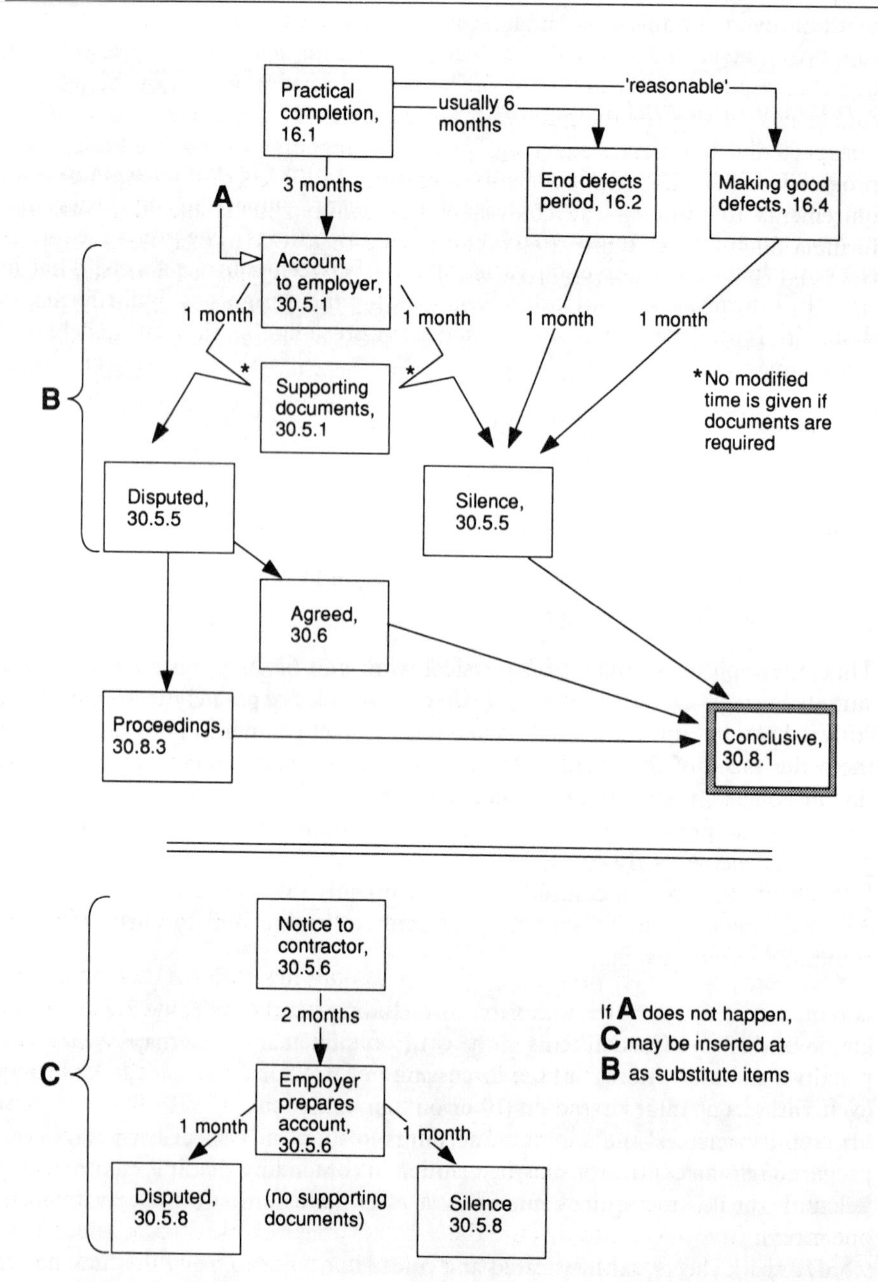

As many of these are predictable, the employer may well be advised to give what he will want in the employer's requirements and whether they should be included in the account or made available separately. A fairly compact account, with supporting detail is far easier to absorb – visually at least! He may also find it useful to stipulate the structure of the account in relation to that of the contract sum analysis, while reserving the right to require changes if the account develops unexpectedly. In practice, much agreement or preparatory work can be achieved progressively, so that stipulations may usefully be added in the employer's requirements to those in the contract about running items like daywork and formula fluctuations. While the employer may thus seek to provoke a response from the contractor, who is often straining at the leash anyway, the employer should not neglect to maintain his own running list of items under (a) above to ensure that he receives them at all, not least those likely to lead to an omission.

The amount of the final account should not be affected by matters like liquidated damages which the contract may give, to use the JCT term, as 'any deductions authorised by the Conditions'. The final account itself is a record of the contract sum adjusted to allow for the works themselves as executed. Equally, it should ignore any outstanding disputes leading to arbitration, although it is prepared 'subject to' these and may be amended as a result.

JCT clause 30(4.1 and parts 4.2 and 4.4) – Payments (Part 2: Final)

This latter episode, clauses 30.5 to 30.9, leaves interim payments and takes the various financial elements into the final account and final statement. However, its sting is in its tail, where it deals with the final statement as closing evidence over the wider satisfaction of contractual obligations. The pattern throughout follows that in JCT contracts dealing with certificates.

Final account

This document, not mentioned in one piece in other JCT forms, is the subject of clauses 30.5.1 to 30.5.3.

The responsibility for preparing the final account lies with the contractor who is required by clause 30.5.1 to submit it, along with the final statement, to the employer 'Within 3 months of Practical Completion'. There is no contractual penalty that the employer can enforce if the contractor overruns this time; not even the second part of the retention depends upon its availability, only any uncertainty over the final balance due. The most that he can do is take over and prepare the account and statement himself under clause 30.5.6, as considered below. If the contractor does submit them, the employer has a minimum time of one month from receipt to dispute the account under clause 30.5.5, which is so worded that delay in submission by the contractor tends to bring the time nearer to this minimum, if he wishes to play it so.

Clause 30.5.1 uses the term 'for agreement by the Employer' of the account. Agreement is not something clearly envisaged under other JCT forms, where documents are hustled to and fro, with the issue of the final certificate as the very next act, to be argued over if necessary. However, the position is not greatly advanced: while the account is 'for agreement', there is no 'agreement to agree' (unenforceable as it would be), only the employer's right to dispute. The effect on the time-scale is taken below.

Upon receiving the account, the employer may require documents reasonably in support of the final account (see list above), even if they are in excess of any laid down in advance, it would appear, so that he can come to agreement or dispute. Clause 12 regarding changes contains no equivalent for the employer of the contractor's 'opportunity to be present' for 'notes and measurements' under other forms, and it might be of limited value in the nature of the contract sum analysis.

Clause 30.5.2 requires the contract sum to be adjusted and the final account set out to see how it is being done. Clause 30.5.3 lists the categories of adjustments, brought together from the four winds and looking very much like those in the other JCT forms (see Table 6.2). Most of them are self-explanatory in the light of discussion of other clauses. A few minor comments are needed:

(a) 'Disbursements' in relation to provisional sum expenditure covers special accounts included direct for fees, specialists and so on.
(b) While the contractor is paid in the account for any clause 22 insurance premiums which he pays on the employer's default, any premiums which the employer pays are recovered by deduction from the account or as a debt.
(c) Other amounts fall outside the account like the last amounts (see Table 6.2).
(d) 'Any other amount' to be deducted or added appears superfluous, as everything required has been covered.

Should the contractor delay in preparing the final account and statement, then as mentioned the employer may step in and do so by clause 30.5.6, preferably after ascertaining that the contractor is not on the verge of disgorging. He cannot act until the end of the three months from practical completion allowed to the contractor, this period not being dependent on whether any making good of defects has occurred. He may then give two months' notice that he will proceed to 'prepare or have [the documents] prepared'. There is no time within which he has to give notice or then to send the documents to the contractor, in fact only in clause 30.5.8 is there even a reference to them as having been sent (not submitted, he is the employer!).

It is possible that the contractor will still come up with his own account before the employer has finalised his (or even at precisely the same time), and there is no provision to debar him from doing this. If he does, the status of the account is open. The rational approach is likely to be to take the contractor's account and use the employer's version so far as it goes to act as a check. Only if the accounts look like being drastically different, may the employer wish to go ahead with his own. If the contractor's account arrives after the employer has sent his, it should be treated as disputing the employer's version by clause 30.5.8, which follows clause 30.5.5 in its approach.

The reference to 'prepare or have prepared' is an unusual recognition of the employer securing outside specialist help, but it does not give him any clear right to include resultant fees in the account.

A major problem for the employer may be lack of data from the contractor. While he, or the specialist acting for him, can assess the value of many changes and so forth, there may be need for invoices or time sheets to deal accurately with some elements. As it is the contractor whose tardiness has led to the problem, it is suggested that a conservative valuation be made, leaving it to the contractor to raise a dispute in due course.

Final statement and payment

This document is mentioned in clause 30.5.1, as being sent to the employer with the final account. Clauses 30.5.4 to 30.5.7 deal with its purely financial aspects, leaving its other aspects to the end of clause 30.

Clause 30.4 gives the content of the final statement as:

(a) 'The amount resulting'; that is, the adjusted contract sum.
(b) 'The sum of amounts'; that is, of all interim payments.
(c) 'The difference (if any) between the two' gives a balance payable, according to the direction of indebtedness.

The item to watch is (b), to ensure that an overpayment is not made. This is possible if 'any deductions authorised by the Conditions', as clause 30.6 terms them, have been made from interim payments. If so, a simple summation of 'amounts already paid' will exclude these amounts, while 'the amount resulting' under (a) will not. The difference will be too high if payable to the contractor. This quirk arises by using the unqualified words 'amounts already paid' in place of 'amounts already stated as due' which in other JCT forms mean amounts *before* deductions are made. The present clause has three possible sets of interim amounts:

(a) The amounts of the contractor's applications under clause 30.3.1.
(b) The amounts paid by the employer in accordance with clause 30.3.3, which are either the same as (a), or are reduced under clause 30.3.4 because the employer's calculations differ.
(c) The net amount actually passing, after reducing (b) under clause 30.4.3 because of any right to do so.

Of these, (a) and (b) both correspond to the 'amounts already stated as due' under other forms. At risk therefore of violating the literal interpretations of clause 30.5.4.2, the calculations should proceed as they do under those forms. An example illustrates this:

	(£)	(£)
Clause 30.5.4.1 'Amount resulting', adjusted contract sum		100,000
Clause 30.5.4.2 'Amounts already paid', actually consisting of:		
Clause 30.3.4 'Amounts properly due'	95,000	
Clause 30.4.3 'Deductions from amounts otherwise due'	2,000 (i)	
	93,000 (ii)	
But, taking the first figure only, *less*		95,000
Balance due		5,000
Clause 30.6 'Deduction authorised', but not yet made		1,000 (iii)
Final amount		£ 4,000 (iv)

All payments, (ii) and (iv), therefore total £97,000. This may also be achieved by deducting (ii) from the adjusted contract sum and then deducting (i) and (iii) from the balance due. This latter method interprets clause 30.6 differently from other JCT versions, as giving '*all* deductions authorised', so that (i) is deducted twice, having been refunded once as well! Such are the problems of vague wording.

The final statement is to be sent to the employer within the same time-scale as the final account, as one leads inevitably to the other. If preparation of the final account reveals any significant amount due to the contractor, and if clause 30.3.1.2 for periodic payments applies, he may submit the account and apply for a further interim payment. He may then be paid before submitting the final statement, which will be affected accordingly. The employer may well not wish to pay the whole balance thrown up by the final account, or indeed any of it, until he has been through the account and he may reduce the payment, or even eliminate it, accordingly. The pattern of occasional interim payments does not cease under the clause on any particular event, other than final payment, but continues as 'further amounts are due'. The contractor may thus apply again during the period for agreeing the account, if the employer discloses some measure of agreement. Interim payment at this stage will therefore necessitate a revision of the final statement, although this is nowhere mentioned. If clause 30.3.1.1 for stage payments applies, the contractor may not receive an interim payment however large the sum thrown up, except that including the second amount of retention if still due.

Clauses 30.5.5 and 30.6 give the timetable following submission to the employer of both the final account and final statement, with the latter governing the position in view of the preceding paragraph. Within these there are the alternative, but comparable provisions already discussed, according to which party prepares the final account and statement. Assuming the more usual situation that the contractor prepares and forwards them, the employer must dispute anything in either document within one month of the end of the defects liability period or of the making good of defects, or within four months of receiving the documents, whichever is the latest of the three. As already indicated, if the contractor delays in submitting the documents long enough, the last of the 'one months' will apply and be very tight.

There is no mention here of the 'supporting documents' of clause 30.5.1. If they were required by the employer's requirements, the final account is incomplete without them and the month runs from their arrival, if later. If the employer 'reasonably require(s)' them when once he has had a chance to examine the account and find his need, the time limit becomes problematic and perhaps open to some partial increase. The larger the omission of supporting documents, the stronger the case for extending the month by the complete period between them being requested and received, but then also the sooner should the employer notice their initial absence and so 'reasonably require' them. This situation is obviously capable of abuse and if this sadly happens the aggrieved party may need arbitration over the 'reasonably' of clause 30.5.1.

The importance of this issue is that if the employer (or, in the alternative situation, the contractor) disputes nothing within the month or other limit of receiving whatever constitutes the complete final account, and the statement, then they become 'conclusive as to the balance due'. Arbitration etc over their contents, as distinct from their other effects in clause 30.8.1, is not possible, so that the uncertainties are particularly regrettable. If, however, the employer does dispute anything in either document, and he could sometimes phrase his requirement of supporting documents as a dispute with a little care, then any time limit apparently goes out of the reckoning and the documents become conclusive when, but only when, 'agreed'. Arbitration before agreement is of course always possible.

Of necessity, there is no time limit placed on agreement, while there is no equivalent to the JCT procedure over an architect's final certificate given in other contracts. That procedure sets time limits over the issue of the certificate, which is not itself dependent on agreement, followed by a short time within which either party may commence arbitration or other proceedings over any disagreement. Usually agreement mollifies all this, but if it comes to it the final procedure, like the prospect of something else even more final, does sharpen the mind wonderfully. In the present contract the parties, with no intervening architect or quantity surveyor, must trundle on as in any such commercial situation, subject to the usual redress not mentioned here, such as action for debt.

While the final account and statement become 'conclusive as to the balance due', by one of the foregoing paths, this balance is still subject to 'any deductions authorised by the Conditions' (see Table 6.2), so that even the direction of indebtedness may be changed and need to be accommodated in the way set out above.

The effect of the final account and final statement
While these two documents are purely financial in content, they may be conclusive over more than payment by the terms of clause 30.8.1, particularly the employer's right to claim liquidated damages (see clause 24.2.1) and satisfaction with some materials and workmanship. If the employer's requirements state that particular materials and workmanship 'are to be to the reasonable satisfaction of the Employer', the position is to that extent similar to that with an architect's final certificate. The agreement or conclusiveness under clause 30.6 of the final account and statement over the balance due is then also conclusive here that the employer is satisfied over quality. He must therefore have satisfied himself carefully while clearing defects, because the contractor's liability ends forthwith, instead of running through the limitation period as with other such defects. If he has stated originally that *all* materials and workmanship are to be to his satisfaction, then the contractor's liability is completely extinguished. This is a relief to the contractor over 'quality' and 'standards', but not over whether, say, an entirely unsuitable material has been specified by him for use in a particular situation. There is by no means an absolute divide between these issues and the parties may find that responsibility is confused in the border area. Even more, it is not a relief over design and the distinction is usually clear here, as design is on the 'far side' of specifying from quality and standards.

This does point up the need for a clear policy in the employer's requirements. the question does not arise later unless the employer qualifies any consent in similar terms. In his requirements, the employer should avoid a blanket mention of 'reasonable satisfaction' like the plague. Only he can carry the consequences, whereas under other JCT forms at least the architect may bear the brunt! Unless there is some pressing need to use the term in a special instance, and this may be doubted, it should not be used at all within the philosophy of this contract. If the employer needs to specify materials and workmanship at all, he should do so in objective terms, such as by reference to British Standards. In general, he should leave these issues to the contractor's proposals or their subsequent evolution. The worst pit he can dig is to say 'You choose everything, but it is all to be to my reasonable satisfaction'!

Any conclusiveness under clause 30.8.1 is subject to two exceptions: fraud and outstanding proceedings. Neither of these exceptions applies to conclusiveness over the balance due, which is settled by clause 30.6. This is a significant difference from other JCT forms. While it is satisfactory over proceedings, because the parties have agreed the amount direct and not with the architect and quantity surveyor intervening, or have had earlier proceedings, it is not so over fraud where the parties would need to rely on their wider legal rights.

Clauses 30.8.2 and 30.8.3 deal with 'arbitration or other proceedings' and again are closely like other such JCT clauses, but are affected by the point just mentioned. Thus clause 30.8.2, in dealing with proceedings 'previously ... commenced', refers to the final account and statement as having 'effect as aforesaid'. While 'aforesaid' is vague, 'effect' can refer back only to where it is previously used, that is, clause 30.8.1. That clause deals only with materials and workmanship and makes conclusiveness over them conditional upon an *achieved* conclusiveness over the amount due in clause 30.6. Any ongoing proceedings before the final account and settlement, make those documents 'subject to' the results (when completed or lapsed) financially. Therefore, until such proceedings are complete and the documents conclusive, their 'effect' over materials and workmanship is suspended.

This is further shown by clause 30.8.3 dealing with 'proceedings ... commenced ... within 14 days after' the two documents 'would otherwise become conclusive' over the subject matter of proceedings. These proceedings curtail the 'effect' of the documents 'as such conclusive evidence' on this matter, so referring back to the 'effect' of clause 30.8.1 again. The '14 days' runs from 'the operation of clause 30.5.5', that is, from the account and statement becoming conclusive in *financial* matters by reason of lack of dispute. There would be little point in them becoming conclusive in that respect, but then being able to be reopened for 14 days. Admittedly there seems little point in them becoming conclusive on that one issue possibly after at least a month's inaction, and it then taking a further period before they become conclusive evidence on other matters, but this is the way it is given.

Clause 30.9 is the usual 'tail ender' that no other payment, corresponding to 'certificate' elsewhere, is conclusive evidence of satisfaction. It adds into its scope 'design', which is surprising as there is no mention of design in any part of clauses 30.5 to 30.8, which is stretching 'aforesaid' to its limit. The position over design remains as at practical completion, which is discussed under clause 16 in relation to defects liability. The contractor's design liability runs on, as would that of an architect and as considered under 'Completion and residual liabilities'.

JCT article 5 and clause 39(9.1) – Arbitration

The arbitration provisions are cast similarly to those in other JCT contracts. They constitute an agreement under the Arbitration Act 1979 to make a reference to a single arbitrator, if there is a dispute. The provisions are not so worded as to prevent a party from unilaterally taking the other to court, instead of going to arbitration by agreement, but comment below on clause 39.5 may be noted. The details of this are governed by the statute. Court proceedings may be better when the issue is principally a point of law; arbitration when matters of fact and technicality are in dispute.

The arbitrator is to be agreed between the parties or be appointed by the person (if any) named in the appendix. It is important to have a name in the appendix and it is common to insert the president of a major professional body. If this is not done, there will be no fall-back position if the parties cannot agree on an arbitrator themselves. It is not desirable to insert a specific arbitrator: he may not be available when needed, or may be unsuited to the subject matter of a particular dispute. This type of contract is potentially open to a very wide range of disputes, from design, through technology and production, into finance. In practice, it appears to be beset by less, due perhaps to more flexibly minded parties, perhaps to uncertainty over an outcome based on the type of contract documentation!

There is no joinder provision, as in some other JCT forms. The purpose of such a provision is to allow sub-contractors to be joined in the same proceedings with the employer and contractor, when the subject matter is essentially the same as would otherwise fall into a separate arbitration between the contractor and sub-contractors. This is obviously economical of effort and more likely to give a direct and equitable conclusion. If such an arrangement is desirable when the contractor simply constructs, it is more so when he designs. It avoids the contractor being piggy-in-the-middle and allows the persons with most to say to do so in the same forum. The parties may wish to introduce such a provision in particular cases.

The scope of arbitration is as wide in article 5 as in the JCT standard form, but is given more briefly in the absence of reference to the action or inaction of an architect. It thus includes the construction of the contract, as well as 'any matter or thing ... arising'. The only matters specifically excluded are statutory tax deduction and value added tax. Clause 39.3 makes the arbitrator's powers subject to the clauses dealing with the validity of the employer's instructions, the effect of the final account and statement, and agreement of fluctuations amounts.

The timing of arbitration is restricted by clause 39.2 on many issues until after practical completion or other ending of work, unless the parties agree to earlier reference. No new arbitration can be commenced later than 14 days after the final account and statement become conclusive, although the comments on clause 30.8 should be noted closely. Any proceedings commenced before then continue, and the conclusiveness is subject to their outcome. The exceptions in clause 39.2 over which either party, as appropriate, may insist on arbitration during progress are matters where postponement would clearly remove much of the effect. They could also produce complete deadlock in the absence of an architect who otherwise acts and so keeps the wheels turning, even if he is later overruled by an arbitrator:

(a) The employer's authority under the contract to issue an instruction on a given matter.
(b) The reasonableness of an employer's instruction that would introduce a change in working obligations or restrictions imposed in the employer's requirements.
(c) Alleged withholding or delaying of a consent or statement by either party.
(d) Extension of time, in any aspect.
(e) Alleged withholding or incorrectness of a payment.

The arbitrator's powers as given in clause 39.3 are expressed very widely, so long as they are exercised over the matter of the dispute. In fact, so far as this aspect goes, the arbitrator may act more widely in what he deals with than may the courts. They are 'without prejudice to the generality of his powers', that is as delineated by statute. Clause 39.4, in giving his award as final and binding, is also affirming the statutory position.

Appeal from an arbitration to the court over a point of law is severely restricted by statute, as is indicated by the extent of clause 39.5. Appeal over matters of fact is in any case not permitted. In essence, the parties must agree to an appeal and the court must be prepared to take it. The clause is a prior agreement to an appeal during a reference or after an award. As such it may be considered to be of too blanket a nature, while the court retains its discretion in all cases. The significance of the point at issue within the arbitration or for the wider law is likely to be the deciding element.

Clause 39.6 deals with 'the proper law of this Contract' and so covers court proceedings as well as arbitration. It gives 'the law of England' as applying, regardless of the nationality etc of persons involved or the location of the works, etc. in question. If any other law is required, this must be stated by amending the clause. It is important that the general law and the arbitration law should go together here. As the present contract is not drafted for use in Scotland, it will be necessary to include a number of other amendments (for instance over insolvency and payment of materials off site), if Scottish law is to apply. A distinct version of the contract covers these points and is published by the Scottish Building Contract Committee. This contains its own articles and appendices, but incorporates the present conditions by reference with a number of amendments that it lists.

There is provision in clause 39.7 for appointment of a replacement arbitrator if the original arbitrator dies or is otherwise lost.

Clause 39.8 provides that the JCT arbitration rules (published separately) are to apply to the conduct of the arbitration. These aim to simplify and speed the process of arbitration by giving procedural rules over many matters and imposing a timetable upon the parties and the arbitrator, which may be extended only with the agreement of the arbitrator. One of three overall procedures is to apply to an individual arbitration:

(a) Procedure without a hearing, with submitted documents only.
(b) Full procedure with a hearing, with representation of the parties and examination of witnesses.
(c) Procedure somewhere between the other two, with a hearing of the parties only and more likelihood of site or other inspection.

Part 3

Alternative approaches

Chapter 11

A comparison of alternative approaches

Some background comments

Tendering, documents and control

Design liability

Contract price

Sub-contractors

This chapter introduces and links the several contractual approaches treated separately in following chapters in this book. The structures of the contract forms related to them are described in Chapter 4. Occasional reference has been made to some of the approaches in preceding chapters, while there is cross-reference between them in later chapters. The approaches fall into three groups for present purposes, although these are taken in a different order in the chapters to ease the detailed descriptive transition from one to another:

(a) Self-contained, supported by free-standing forms of main contract:
  Chapter 13: BPF/ACA Form of building agreement.
  Chapter 15: GC/Works/1 Single-stage design and build contract.
  Chapter 16: ICE Design and construct contract.

(b) Modifications to other approaches, supported by supplements to main contracts:
  Chapter 12: JCT Designed portion supplement to the form with quantities.
  Chapter 14: JCT/BPF Supplement to the form with contractor's design.

(c) Subsidiary parts of other approaches, contained within main contracts:
  Chapter 17: JCT clauses for performance specified work in the forms with quantities etc.
  Chapter 18: JCT nominated sub-contract to the forms with quantities etc (corresponding, less detailed provisions occur in other contracts).

There are several potentially common features shared between, differing between or omitted from some of these contractual approaches which are most easily highlighted by being brought together here, using abbreviations of the above (already abbreviated) titles. The JCT system of nominated sub-contracts is discussed in Chapter 18, as this is the most detailed of such sub-contract systems. For present purposes other such systems, which exist in relation to some of the main contracts, are ignored and may be assumed to be similar within these limits.

The notes preceding the GC/Works/1 contract state that it is best used for 'relatively simple buildings'. The BPF/ACA form is intended for 'medium sized' buildings. The other documents are not expressly limited, although the JCT designed portion supplement, performance specified work clauses and nominated sub-contract all occur as subsidiary arrangements, with at least some limit of scale according to the main contract in question. The ICE contract differs from the rest by not being for building work at all. It is included in this book because of its particular approaches for comparison purposes.

First a few wider comments are given to point up parts of the discussion and to hark back to elements of what has been raised in Part 2 of this book. The comments are simply pointers and should not be read alone to obtain a proper view on the matters mentioned.

Some background comments

The several stages set out by the RIBA Plan of Work introduced in Chapter 5 show a shift of controlling activity through the life of a project. In general terms, the client is dominant during inception and feasibility, the designer from outline proposals to production information (which may be taken as including bills of quantities or their equivalent) and the constructor from tender action to completion. These stages tend to overlap or be subject to reiteration, but broadly the emphasis holds. In particular, the client is more distant from post-contract design and construction and cannot influence their process or progress as much as he can those of pre-contract design. If he does, there may be a greater cost penalty to meet. More aware clients may seek to strengthen their role alongside the designer or constructor or both and this aim may be seen in aspects of the approaches grouped under (a) above and also in the JCT/BPF supplement.

In design and build as considered in Part 2 of this book, the main change from traditional working is that more of the design operation often comes alongside and is more closely related to the construction and also is more remote from the client. This shows in the pattern of precise detailed decision-making; this largely moves from the client over design when the contract is placed, unless he expresses himself in fairly open terms by changing his requirements. There is also a deliberate postponement of some decision-making by the contractor into the post-contract period, as against what is usually intended in the traditional arrangements which separate design and construction responsibilities. Under the strict JCT pattern, leaving aside the supplement and other contracts, there are limited formal opportunities for the employer to inspect drawings and other information in a reactive way ahead of construction so that a measure of tacit approval or monitoring can be exercised.

Another aspect of 'control' is the extent and nature of tendering, taken further under the next heading. It is usually the case that the tendering list for design and build is shorter than that for a comparable project procured by traditional means.

This reflects the greater area of responsibility, and so of work, assumed by the contractor at that stage. To keep this work within reasonable bounds and also to save overall project time, the feature of more extensive post-contract design again becomes significant. Only two approaches make any statement over tendering and tend to go against this concept. The BPF/ACA system in effect lays down that the contractor is to develop design detail either before or after tendering, but in either case at his own risk as to its adequacy and without price adjustment. The GC/Works/1 contract is a single-stage design and tender with complete design prepared at the tender stage, although not necessarily fully approved by the client.

Against all of this, there is the compensating question of design liability assumed by the contractor, this being at the level of that of an independent professional designer under most arrangements. This is one of the various elements used to divide out the risks inherent in the project, while devices extraneous to the contract such as collateral warranties may be invoked to reapportion or spread the risks more widely.

While the JCT with contractor's design form transfers the total design of the project to the contractor, subject to what the client has already done and for which he may choose to retain responsibility, there is the alternative of passing to the contractor only part of the design. This may be done by restricting his tasks at the initial stages so that he comes in later, perhaps just to complete detailing of the whole. The BPF/ACA contract goes along this route, but leaves design liabilities vague and potentially divided. The JCT designed portion supplement, performance specified work and nominated sub-contract approaches cut the cake differently by giving the contractor or his sub-contractors just part of the project to design, within a framework in which the client's designers perform and remain responsible for the overall design in outline and some parts of it in detail. Any of these approaches is something less than a full design and build system, even though the stage or element may be carried out under similar principles.

When once part only of the design responsibility passes to the contractor, the field is set for the possibility of split liability for design faults and for the certain need to co-ordinate and integrate the various design contributions on behalf of the client.

Some of these issues have received consideration in varying levels of detail in earlier chapters. They recur where relevant in those that follow and, with others, are highlighted in the rest of this chapter.

Tendering, documents and control

Tendering and documents

The various contracts are silent in the main about the methods of tendering which are compatible with them. While negotiation rather than competition is always possible, there is also the quite distinct matter of single- or two-stage tendering.

Only the notes introducing the GC/Works/1 contract are specific by ruling out two-stage design. Even this does not necessarily exclude two-stage *tendering* in the sense that it is generally used in this book. It refers to the special case of a second stage of *design* to be completed after acceptance of the tender and before construction starts and states that this would require special provisions. Presumably the arrangement, like the single-stage document discussed here, would still carry provisions about design submission and monitoring, so that design completion would remain a relative term.

Although the JCT designed portion supplement is also silent, it is not very practicable to have the main part of the project subject to a single-stage tender based on bills of quantities, while having a two-stage tendering system for the portion of the project which is covered by the supplement. This is even more true of the smaller-scale option of JCT performance specified work. Beyond these cases, nothing is stated or may be inferred in the other options about tendering.

Over discrepancies, and so over general precedence of design documents, the GC/Works/1 and ICE contracts give the requirements (variously of the authority and the employer) as prevailing over what the contractor puts forward. This reverses the JCT position and puts the contractor on notice to check that he does not change, as distinct from develop, the requirements by his design or specification. In the BPF/ACA form, the main design is by the client's designers and the contractor assumes a design responsibility only for the parts which he designs (but see the comments in Ch. 13 on his warranty over meeting performance specification). There are no proposals at the time of tendering, just 'necessary' drawings and details put forward during progress, which effectively are subsidiary to the client's required design. All the JCT cases allocate responsibility by following a similar line to that of the contractor's design contract.

Control

A strong control pattern is given in the main contracts outside the 'straight' JCT set assumed for most discussion in Part 2 of this book. All four, including here the JCT/BPF supplement, provide for the supply of design information to the client or his representative, with them having powers to query and seek modifications to overcome not only divergences from the requirements or proposals (or their equivalents), but also misgivings about how technically satisfactory they might be. These provisions are drawn to leave the contractor responsible for any design which is still defective. They all require the contractor to furnish quotations, either automatically or when so required, for all variations and loss and expense or similar before work proceeds, with fall-back provisions if agreement cannot then be reached.

Again, all of them have provisions for dispute resolution ahead of arbitration and during progress, variously referring to this as adjudication or conciliation. To an extent these soften the harshness of the client orientation of the contracts, but they also press the parties more strongly into reaching agreement, as otherwise they lead more irrevocably towards arbitration.

The GC/Works/1 and BPF/ACA contracts go further and emphasise, among other matters, the power of the project manager or the client's representative respectively to give wide-ranging instructions and check into the programme and if necessary seek revisions or acceleration. They also bring the client more actively into making decisions and into the general action. The JCT/BPF supplement follows this pattern in some respects, to give a part-way arrangement.

The ICE contract is distinctly open in these areas, saying less that can be read specifically in either direction. This is standard civil engineering practice, in which a great deal depends on tacit acceptance of the engineer's quite sweeping authority, counterbalanced by reimbursement when this leads to the unexpected. Since civil engineering is regularly conducted against the broader background of the unexpected in terms also of natural hazards, this appears to be a reasonable overall contractual philosophy accepted as the norm by all the participants. Even clients usually go along with this, if only because the individual causes of higher expenditure cannot always be distinguished easily.

Design liability

All the JCT group of documents stay with reasonable care as the criterion over level of liability for design faults, as might be expected. This too is the criterion in the GC/Works/1 contract and mainly in the ICE contract. The GC/Works/1, ICE and BPF/ACA contracts have provisions about groundworks which should reasonably be foreseen variously by 'an experienced contractor' or equivalent, which in some circumstances could be read as overlapping with design responsibility to produce a strict care standard. There are further provisions in the BPF/ACA form covering a warranty about performance specifications which carry a strict care obligation and covering the fitness for purpose of what the contractor designs and perhaps of the wider works.

The BPF/ACA and GC/Works/1 contracts both require the contractor to maintain design indemnity insurance, whereas all the other arrangements are silent on the point.

Contract price

The various approaches all rely on some form of lump price, as does the JCT with contractor's design contract. However, this varies with the particular nature of the contract. The JCT/BPF supplement does not change the pattern of its parent form, although it does allow for the contract sum analysis to consist of bills of quantities, or conceivably to contain some element of quantities. Quantities as a basis are discussed critically in Chapter 9.

The BPF/ACA form provides for a schedule of activities (the standard document in the BPF system) or for bills of quantities. The GC/Works/1 contract has a pricing document of unspecified form, so paralleling the JCT contract. The ICE contract is unclear on the whole issue of contract price; it *appears* to rely on a lump sum adjustable on a similar basis to the JCT contract. It may contain approximate quantities for parts of the works, which is understandable in principle in view of the many uncertainties inherent in civil engineering work, if difficult to handle in practice when the contractor designs.

The JCT nominated sub-contract is a reduced version of a main contract in price terms and may be based on the various patterns of a plain lump sum, firm quantities or approximate quantities, without regard to who performs the design.

The JCT designed portion supplement is a design and build lump sum accommodated within an otherwise standard JCT contract with quantities. It has its own analysis, with the form of this left open as under the JCT with contractor's design contract. The performance specified work arrangement is also within such a contract and itself consists of a set of quantities, or perhaps a provisional sum, within the general bills.

All of these follow the regular method of adjusting for changes or variations appropriate to them. Even the ICE contract provides for this, despite its coyness over the completeness and inviolability of the contract sum. As noted under 'Tendering, documents and control' above, the JCT/BPF supplement and the three main contracts provide for quotations to be given for hopeful agreement in advance of performing varied work, with these covering loss and expense or its equivalent in the particular contract.

Provisional sums in the employer's requirements are already allowable for the JCT/BPF supplement, by virtue of its main document. They occur in the ICE contract, but without clarification of who includes them and why, an acceptable position if they are simply the equivalent of approximate quantities and so stemming from the client's side. Neither the BPF/ACA nor the GC/Works/1 contract provides for provisional sums, although the notes preceding the latter suggest that, if exceptionally needed, they should be included in a way reminiscent of the JCT approach.

Sub-contractors

In general the various documents allow domestic sub-contracting or sub-sub-contracting with consent at main contract level from the client's side. Only the ICE contract allows for 'Prime Cost Items', although it does not differentiate between them and 'Contingencies' also mentioned (the latter meaning 'provisional sums'). There are no terms for nominated sub-contracts as such.

The ICE contract limits its mention of sub-contracting to that for construction, even if in rather negative terms, while requiring appendix entries naming design consultants to be made if known at the time of entering into the contract. If not, the work concerned is to be identified. All the other main contracts specifically deal with design sub-contracting as during progress.

A named sub-contract arrangement in the sub-contract documents is introduced by the JCT/BPF supplement and the BPF/ACA form, these being strongly related documents here. The former proceeds along the lines of the JCT intermediate form, as might be expected, with procedural and other detail. The latter simply requires the contractor to sub-let to suppliers or sub-contractors named, without spelling out conditions.

Chapter 12

JCT designed portion supplement

Pre-contract design

Post-contract design

Financial provisions

This chapter is devoted to the tailor-made JCT contractor's designed portion supplement, introduced in Chapter 4 and to which JCT Practice Note CD/2 applies. The supplement is intended for a relatively large, fairly self-contained part of the works, such as a distinct structure or functioning element, as is described further under the first heading below. An alternative approach intended for smaller elements of work, usually less independent in character, is to use the clauses for performance specified work (at one time viewed as contractor-designed construction by standard methods of measurement) contained in the various editions of the JCT standard form of building contract. This approach is considered in Chapter 17, where the distinctions between the clauses and the present supplement are outlined. For suitable projects, both the supplement and the clauses might be used within the same contract.

Chapters in Part 2 have discussed all clauses in the JCT with contractor's design building contract, and that discussion is assumed here, usually without any reference to it. These earlier chapters make frequent comparisons with the JCT standard form of building contract, mostly without differentiating between the six editions of the contract. The supplement, as it is termed here, is essentially a list of modifications of two of these six: the private and local authorities editions, both with quantities. In this chapter, therefore, the same extent of familiarity by the reader with these forms is assumed and comment centres on comparing the use and effect of the supplement with those of the with-design form. The clause numbers of the with-quantities form differ slightly. These are the numbers referred to in this chapter and in the form as modified by the supplement. Only a few clauses warrant comment and these are taken below in the context of topics which mostly follow the broader sequence adopted in Part 2.

Reference is not made here to the revised clause numbers of the anticipated JCT section headed contract (see the comparison table in the Index of JCT clauses).

Pre-contract design

Scope

The supplement is intended to allow the design of some part of a project to be delegated to the contractor within the framework of the architect's overall design, which includes that of other consultants of the employer. Its primary effect is to graft into an otherwise 'build-only' contract the special provisions about the contractor's design contained in the design and build contract, so that they apply to the designed portion only. This is therefore done without resort to the concept of nominated sub-contracting, discussed in Chapter 18 as a means of securing design from the construction side, although nomination may well be used within the part of the works outside the designed portion.

Within this framework, the architect proceeds with the total design, completing part of it and developing what will become the contractor-designed portion so far as desirable to relate it to the rest, and then preparing a set of employer's requirements, as in a full design and build case. The contractor tenders for both parts together, basing the one on bills of quantities supplied to him and forming part of the contract, and the other upon a set of contractor's proposals of his own devising. The latter are supported by 'the Analysis' which is the equivalent of the contract sum analysis, but which covers only the designed portion. The whole of the works goes on together, with each part subject to its own procedures, but with these interlocking in the one contract. The architect co-ordinates the design of the portion with the rest, although the contractor develops it and remains responsible for it. In so doing, the architect effectively adopts the role of employer's agent over matters of inspection etc, as in the with-design form, but he does not have the additional powers which the employer may choose there to delegate to the agent proper under article 3 (see Ch. 7), as he is not designated agent.

As indicated in Chapters 2 and 4, this approach is suited to various components of the scheme, provided they are distinct entities with their own design criteria, so that design responsibilities are not confused. Separate buildings or complete superstructures are strong candidates. So too are elements otherwise amenable to nomination, such as structural and services work, and also lifts and other work localised within a building. In general a designed portion is useful to secure specialised design, while there is little advantage in using it for work which is similar to other work which the architect is designing. It is better in the latter case to have design by architect or contractor, but not both.

Several questions arise naturally out of this outline, which are taken during the comments below.

Briefing and competition

There are two crucial aspects about briefing: its timing and integration with the rest of the design. In the absence of competition, these are comparatively straightforward to deal with, even though care over the detail is essential. Competition between tenderers complicates matters, as they are tendering for a scheme partly common in detail and partly not.

While therefore the patterns described in Chapter 5 over briefing and in Chapter 6 over competition may be kept to, the way in which they are used will vary according to the designed portion. If this is free-standing within the scheme, quite early briefing is likely to be practicable. This will allow the two parts of the design to develop together, and each to take account of the other. The main problem may be how the architect's part can take account of several diverging designed portions at once. There are three broad options:

(a) To pull all the designs into conformity with the architect's part as a core element.
(b) To eliminate several designs, perhaps all but one, in a first stage of competition, so reducing or removing competition in the second stage.
(c) To allow the designs to go ahead, with any minimum degree of guidance, and be prepared to adjust the architect's design and the bills of quantities, after tendering but before or after entering into the contract.

Which to use is a matter of priorities and policy, conditioned by how far the designed portion dominates the scheme as a whole. The last option bears on the question of firm or approximate quantities. Usually the design uncertainties in an approximate quantities contract militate against the contractor being able to design something firm pre-tender, because he cannot relate to an existing firm design. Sometimes that other design cannot be made firm until the contractor has designed his portion and here the supplement approach could be very useful. In the case of a without quantities contract, the intention is that the works are small, so that normally there is not room for two designers. It is necessary in any case to supply quite an amount of information in the formal employer's requirements, so that the contractor is not put at undue risk over the question of design integration discussed below.

If the designed portion integrates closely with the rest, the desirable timing for briefing is more problematic. The portion may be determinate of the rest, if it is something major like the structural system or complex services. If so, early briefing is essential, with the potential intensification of the problem of choosing between the three options listed above. If the portion is subsidiary or contained within a fairly small section of the works, it may be possible to defer briefing until much later, as there is less design to be performed and it will have fewer repercussions.

Inherent in early briefing is the need to secure as definite a commitment as possible from tenderers that they will not withdraw before tendering, because of the commercial and technical disruption this would cause. This is more difficult when the designed portion is only a small part of the whole, as tenderers are being asked to forgo commercial flexibility over future work, when they have made only a relatively small investment of design effort for the scale of the potential work.

Table 12.1 Design provisions in contractor's designed portion supplement set out to compare, where possible, with those in the JCT with-design form

Clause etc	*Subject*
Seventh recital	Repetition of 'satisfied that they appear to meet' statement
2.1.3	Contractor obliged to complete the design, with requirement about architect's directions over integration
2.2 to 2.4	Provisions about errors, discrepancies and divergences
2.5	Contractor to provide architect with design information
2.6	Repetition of design warranty and liability provisions
2.7	Safeguards to contractor over architect's directions about design integration
2.8	Architect may notify contractor of apparent design defects
2.9	Position qualified over extension of time, loss and expense and determination
5.8	Contractor obliged to provide as-built and other information
6.1	Provisions about statutory obligations extended
6.2.4	Statutory fees and charges for the portion included in contract sum
8.1	Contractor restricted from substituting materials etc in the portion without architect's consent
27.6.6	Contractor obliged to provide drawings etc on determination
28.4.4	Contractor obliged to provide drawings etc on determination
28A.3.2	Contractor obliged to provide drawings etc on determination
30.10	Portion excluded from normal scope of architect's final certificate; to be read in conjunction with clauses 2.1.4 and 30.9.1.1

Post-contract design

The supplement introduces a number of main elements over design, partly as in the design and build contract, to apply to the designed portion only. These provisions are summarised in Table 12.1 and give a number of identical results to those in the with-design form, but there are also some explicit or implicit shades of difference. They are far more extensive in their effects than the provisions for performance specified work (see Ch. 17).

Integration

At the point of accepting the tender, the employer is adopting the same stance over the designed portion as over a full scheme: that of it 'appearing to meet'. This has to take account of how the portion relates to the architect's design for the rest of the project. So far as the portion is already designed, the employer must be held to be fully aware of this relationship. For a complete building, this will cover its massing and positioning in relation to the rest of the scheme and other visual aspects, leaving many other and more technical aspects as only 'appearing to meet' the requirements, until shown to do so in the finished building. For an element of a building, the corresponding visual aspects must be fully accepted by the employer, but not all the technical aspects again.

In each case, there are some matters where the employer must be held to have accepted the technicalities, at least at the general level, such as the broad user differences between ducted warm air heating and underfloor heating. In the case of an element, the architect on behalf of the employer is bound to look more closely, because of the way in which it is to be integrated into the rest of the design for which he is responsible. Desirably, performance specified work within the contract bills will have been used instead when particularly close control of the physical disposition and dimensions of the element is needed, so that this is achieved in advance and the presentation of the equivalent of contractor's proposals is postponed to the post-contract phase.

Here the supplementation of article 1 and clause 2.1 (by means of clause 2.1.3) becomes important in the post-contract phase. These both provide in similar terms for the design of the portion to be completed, the clause having 'comply with the directions' of the architect 'for the integration of the design with the design for the Works as a whole'. For most purposes under the JCT standard contract, the architect gives instructions, which the contractor is obliged to obey, subject to certain precautions. The term 'directions' indicates that the present action is not so strong. It should be couched in terms similar to those requiring a change under the design and build contract, that is, setting out what end is to be met, so that the contractor may put forward his own design solution. The architect must not enter into the direct design of the portion. The purpose of his directions is to secure integration of the design, not to determine it. There is obviously an uncertain area between directing integration, which must involve marginally either changing design or developing it differently from what had been intended, and instructing a variation (the term for a change under the supplement) as a definite modification of the employer's requirements.

It may be argued that a variation modifies the employer's requirements and so comes from the employer, whereas a direction over integration simply secures that the employer's requirements are achieved and so comes from the architect. This is true enough, but there is the further distinction that a variation leads to an adjustment of the contract sum, whereas a direction does not. It is also common for the architect under any contract to instruct variations without the immediate authority of the employer. This is accepted practice to secure practical working for essentially technical changes that can be accommodated within some margin, although this is not often defined. It is not at all unknown for it to be exceeded, on any reasonable estimate of the putative margin. None of this is of any concern to the contractor, apart from whether he will be paid or not, or concede a deduction.

The contractor must be prepared to accept some incidental change, such as repositioning ventilators in a ductwork installation, provided that his design is not compromised and that he is directed before any affected physical work has been performed. He should be prepared to accept minor effects, if these are within whatever may be termed 'the broad philosophy of his proposals', meaning a bit of give and take in the interests of goodwill. He has a right under clause 2.7 to 'specify (any) injurious affection' to 'the efficacy of his design', and only has to proceed after the architect has confirmed his direction. This gives an implied protection over liability, but does nothing about whether reimbursement is possible (see later comment).

Beyond this he may seek a variation instruction and so payment, or, if refused, arbitration, although not during progress. If his design is seriously undeveloped when he tenders, he should review the possible areas for major 'integration' carefully before entering into the contract and seek written clarification from the architect of the limits of what he is going to require. These may lead to a revision of the tender but should, in any case, be incorporated into the contract as a sort of 'employer's proposals'.

Specification

Within the regulating framework of any post-contract integration by the architect, the contractor remains entirely responsible for his design. Thus clause 2.1.3, containing the integration provisions, is otherwise the requirement for the contractor to 'complete the design', including any specification so far as not already in the employer's requirements or the contractor's proposals. This is just as in the design and build clause and effectively what happens to performance specified work.

In view of the split in responsibility, this specifying of work and materials needs care. If some materials are mandatory in the contractor-designed portion or if there is a restricted choice, it will be tempting where possible to say that these and their workmanship are to be 'as described in the Contract Bills', that is, in the architect-designed portion. This can be a useful shorthand, but may lead to a division of responsibility if a failure could be attributed to either design or specification. It may be suitable to hedge the position about with provisions over the contractor taking on the architect's specification, and taking all responsibility with it. This could be read as 'use it sensibly' or 'I don't trust what I have written'! It may be far better to invite the contractor to put forward his own specification, or to say, when he so wishes, that the architect's specification is incorporated. If finishes etc in the two parts of the scheme are to match, that much should always be said, even if fixing or application is allowed to diverge.

A subsidiary point over specification occurs by reason of clause 2.1.4, which exists in the JCT standard contract as part of the superseded clause 2.1. This provides that where qualities and standards are to be at the architect's approval, this shall be reasonable. The effect of clause 30.9.1.1 is that the architect remains liable for any defects in such cases after issue of the final certificate, and not the contractor. A blanket clause that 'materials and workmanship not otherwise described shall be to the architect's approval' is always a time bomb in this respect. It is the more so if the contractor chooses some such element within the designed portion, so making the architect responsible for what he has not chosen and possibly would not have chosen, had he known.

Development

Discussion of the overarching integration of design has bypassed the related, more basic question of design responsibility. This also is given in clause 2.1.3 and reproduces the effect of the with-design form about completing the design, including specifying materials and workmanship as just discussed. Several provisions follow about defects or clashes in or between documents, and these differ in part, because of the mixture of origins that documents possess.

Errors etc

If there is an 'error in description or in quantity . . . or . . . an omission of items' in the contractor's proposals or the analysis, but no problem elsewhere, clause 2.2.2.3 provides that there is to be no addition to the contract sum for any direct correction, or for any variation to the remainder of the works due entirely to this correction. The intention is clear, although its application is more difficult:

(a) In the cases of both contractor's proposals and analysis, these support the contractor's lump price and no internal adjustment is to be expected, following the usual package principles.
(b) In the case of the contractor's proposals, so long as these are put forward in relation to precise design for the architect-designed portion, it is again reasonable that the contractor should absorb the effect of 'error in description'.
(c) If there is an 'omission of items', an extreme form of 'error in quantity', it is likewise reasonable not to make up the deficiency from the architect-designed portion.
(d) As the analysis exists only for limited uses, it is difficult to see how correction can affect the contract sum: at most it should distribute figures within the same total, on the principles set out in Chapter 9.

The other question is whether there should be any deduction from the contract sum:

(a) Within the analysis, (d) above is the applicable principle, so that no deduction arises.
(b) Beyond the contractor's proposals and analysis, it is difficult to envisage how a deduction can be possible, if design there is precise and the portion has been tailored to fit.
(c) If the correction does edge the portion into the rest in some way, however, a deduction resulting from a normal variation omission would appear reasonable.
(d) If there are both additions and deductions, not necessarily of the same items,

it is reasonable to offset one against the other before applying these suggested rules.

If the architect-designed portion is not absolutely defined at the date of tender, so that either there are approximate quantities or variations are inevitable, these two sets of principles (a) to (d) are still relevant. It will be necessary to measure what would have been needed if no adjustment of the portion had arisen, and then use this as the benchmark for deciding the rest.

Clause 2.3 deals with 'any discrepancy within or divergence between' the whole range of pre-contract and post-contract documents emanating from either the architect or the contractor, with the exception of the conditions etc themselves, which prevail by virtue of clause 2.2.1. The clause considers only the contractor finding a problem, so that the architect issues an instruction to resolve it, or provoke its resolution if this needs the contractor's design. It is to be hoped that the architect will deal similarly over anything he finds himself.

Clause 2.4.1 acts as a rider to clause 2.3 and deals with 'discrepancy or divergence' entirely within documents produced by the contractor, at whatever stage. Here the contractor is to state as soon as practicable his 'proposed amendments' and the architect does not have to instruct until this is done. This allows him to evaluate the proposal and perhaps integrate it. His instruction, or any prior reference back, should avoid assuming responsibility for any design, as usual. If a divergence is between documents of the architect and contractor, and the architect requires amendment of the contractor's documents, it is reasonable that the clause 2.4.1 procedure should apply there as well, although it is not explicitly covered.

If resolution of the problem falls under clause 2.4.1, clause 2.4.2 provides that there shall be no addition to the contract sum, and this should follow for the cognate amendment just mentioned. The possibility of a deduction is not mentioned but, if alterations spill over into the rest of the works, one might arise. Matters are more likely to be self-contained and no deduction should then be sought. When the problem lies within architect-produced documents, the usual rules for valuing the effect of an instruction will follow.

These clauses may be compared with the more simple clauses 2.3 and 2.4 in the with-design form. In effect, each party chooses which to accept out of discrepant options in the other's documents, while the contractor's proposals override the employer's requirements by virtue of the 'appear to meet' provision. In none of these cases is the contract sum adjusted, unless the employer is dissatisfied with the impending physical result and instructs a change.

Responsibility

This issue is embraced by clauses 2.5 to 2.9. Of these, clause 2.6 is a direct repeat of clause 2.5 in the with-design form, subject to drafting amendments, and needs no special comment. This contains the central statement over possible liability. The other clauses arise mainly out of the shared design situation.

Clause 2.5.1 repeats the requirements of clause 5.4 of the design form for the contractor to supply copies of drawings, details and specifications. To these are added '(if requested) calculations', while the whole list is qualified as 'to explain or amplify the Contractor's Proposals', so balancing the reason for the architect supplying similar information, but without mentioning that the contractor 'uses (them) for the purpose of' performing the designed portion. 'Levels and setting out dimensions' are given separately and this mention *is* made of them. These points are a significant shift of emphasis, underlined in clause 2.5.2, which requires the contractor to hold off 'any work to which these copies relate' for at least fourteen days after delivery to the architect. They link with clause 2.8, which requires the architect to 'give notice to the Contractor' of anything 'which appears' to him to be a design defect. The clause continues, however, that 'no such notice or want of notice' is 'to relieve the Contractor of his obligations'.

The effect of clauses 2.5 and 2.8 is thus to give the architect the opportunity to check the contractor's design before it is put into action, in as much detail as he wishes. This he may overdo, but the facility is very useful to him when the two parts of design, his and the contractor's, entwine significantly. It is not to be construed that, by making no comment, he is giving positive, if tacit, approval. He *must* give notice of apparent defects that he finds, but he does not relieve the contractor by default, or if he does give notice and the contractor presumably acts upon the criticism. It may be doubted whether, in an extreme instance, this would be sufficient to relieve the *architect* when he might have a duty of care. This might extend towards the contractor and is very likely to exist towards the employer. The architect should therefore ensure that his terms of engagement are in the standard RIBA form at this point, over delegation of design responsibility.

The contractor, on the other hand, is not obliged by clause 2.8 to do anything about the architect's notice of defect, and the architect has no further power over him, other than to instruct a variation requiring the contractor to modify his design. This must be distinguished from a direction over integration as previously discussed. Thus a direction to amend the profile of a concrete beam for visual or functional reasons might fall within integration, while having pronounced structural effects, whereas a communication about the steel reinforcement of the beam certainly would not be integration. Re-routing a pipe could well be integration; changing its size might serve the same purpose, but might need some other change as well (say of a radiator or of insulation) to keep the presumed design efficient.

Because of the wide possibilities inherent in this whole area, clause 2.7 already mentioned gives the contractor the right to give notice to the architect over what 'injuriously affects the efficacy of the design' for which he is responsible, as happens also with performance specified work. This refers specifically to directions over integration and generally to instructions. If the architect comments unfavourably on a design matter and the contractor does not change it, but the architect instructs a variation, the contractor then has this position on which to fall back. Even so, if the architect confirms (that is, repeats) his direction or instruction, the contractor must then comply. It must be inferred that responsibility for a matter that is aesthetic, technical or anything else, then passes from the contractor to the architect. It is required that the contractor shall give his dissenting notice within seven days of hearing from the architect, although, when time permits, it seems reasonable to countenance longer.

While clause 2.7 appears to protect the contractor over liability, it does mean that the contractor must be very explicit over any highly specialised matter. He may be advised to give a 'holding objection', pending full details from a consultant or sub-contractor, which may take more than seven days to seek and obtain.

It is also the case that the clause affords protection, but not reimbursement, if the contractor happens to be forced into greater expense than he intended and cannot squeeze a variation out of the situation. He may of course incur less. The situation is analogous to that in which the contractor has to amend his design, and so his expenditure, over a planning decision. The analogy fails in that the architect is an 'insider' as a planning officer is not, and the contractor may consider him an insider committed to a position in advance. Both contractor and architect should view matters in the spirit of the contract, made more difficult by the interaction of two designs. In the end the contractor must conform if needs be, subject to the possibility of arbitration over reimbursement after practical completion, when the arbitrator is likely to deserve everyone's sympathy.

The group of clauses ends with clause 2.9 which covers that the contractor cannot seek extension of time, loss and expense or determination, when progress has been affected by his own mistake in the documents or lateness in providing information. All of this refers to the works as a whole, and not just to the portion.

Contingent matters

Later clauses contain predictable provisions about design, as shown by Table 12.1. One is clause 19.2.2 which controls sub-letting of design of the portion or, presumably, any part of it. Here it would be reasonable for the employer to withhold consent to a proposal to introduce one of his own consultants, especially his architect, into the contractor's team. While this would no doubt ease integration, it could lead to a conflict of interests, as the JCT practice note recognises. The same is possible if the employer seeks a common team in his requirements, and the contractor should resist what could become a Trojan horse if relations deteriorate. The possibility of the employer giving a list of at least three names, from which the contractor may choose when sub-letting, is present here by virtue of clause 19.3.

Not predictable is the omission from clause 28 of a requirement for the contractor to provide drawings etc for the employer's retention when he determines against the employer. This occurs under the main design and build form, and should be introduced here also, otherwise the employer may have difficulty if he wishes to complete the works. Clause 30.10, over the conclusiveness of the architect's final certificate, segregates work and materials as a whole from design of the portion in particular. As discussed over the main form in Chapter 10, this should leave design liability with the contractor, unless something untoward has been done. Liability over work and materials also remains with him, subject to the 'reasonable satisfaction' question under clauses 2.1.4 and 30.9.1.1.

Financial provisions

In between the numerous drafting details in the supplement needing no comment, there are three aspects of payments to be highlighted.

Clause 13 dealing with variations and provisional sums contains no procedure for the contractor giving his consent to any variation to the designed portion, as he does under the with-design form. If the portion is closely bound in with the rest of the works, this is likely to be the best approach. If it is not, the omission may deprive the contractor of a significant right when his design is crucially important, so that he may be advised to seek a modification before entering into the contract. Positively, clause 13.2 allows the architect to issue variation instructions: in the case of the portion, these must be statements about changing the employer's requirements so that the design solution is left to the contractor. Several other parts of clause 13 amplify the rules over valuation to cover use of the analysis and the element of design.

Those parts of clause 30 dealing with payments contain no substantive points. There is not a set of provisions dealing with stage payments, which would normally be awkward to operate in a contract otherwise based upon periodic payments. Clause 30.2 does have the words 'subject to any agreement between the parties as to stage payments', but no further details are given. In a suitable case (such as when the bulk of the works were covered by the supplement), the provisions of the with-design form could be hung upon this peg, using any of the procedures discussed in Chapter 10. It might be best to arrange that the whole of the works should then be paid for by stages.

Clause 40 over formula fluctuations contains a number of modifications to the with-quantities formula rules, which otherwise are given separately from the contract but incorporated by reference. There is thus one consolidated set of rules referring to the works as a whole, with subsidiary provisions for each portion, rather than a separate set for each. Among the rules, rule 11e states that the parties are deemed to have agreed the divisions of the analysis for fluctuations purposes only, and rule 11f requires the quantity surveyor to include them as put forward 'when submitting the contractor's proposals'. This is better than the post-contract amplification permitted under the with-design form.

Chapter 13

BPF system and BPF/ACA agreement

Some features of the system

Documents under the agreement

Contractor's design responsibilities

Design-related provisions

Financial provisions

This chapter deals with the elements of a system for procuring construction work which differs in a number of significant ways from that related to the JCT family of contracts. The system is set out in the 'Manual of the BPF System', which is not a form of contract, but a guide to how the British Property Federation system may be used both pre- and post-contractually, subject to whatever modifications a particular client may wish to make. A particular edition of the Association of Consultant Architects' Form of Building Agreement is issued as a contract specifically to suit the system. These two documents are introduced in Chapter 4 and form the basis of the discussion in this chapter. They are referred to throughout this book as the BPF system and the BPF/ACA agreement or form. The BPF/JCT supplement taken in Chapter 14 acts as an alternative which suits some aspects of the BPF system, while not being restricted to use with that system.

Consideration in this chapter focuses on the design aspects of both system and contract, with limited mention only of other important issues, such as financial matters. The BPF system was innovative when introduced and has received a moderate level of practical use, but it contains quite a number of features of interest for comparative purposes. The system is not explained here as a whole and it is necessary to study the manual to obtain a rounded view, but some key points are summarised before considering the aspects related closely to the agreement.

The system is intended to encourage greater co-operation between the several participants in the construction process, including the client, by setting out more precisely than is often the case who is responsible for doing what and, critically, when they have to act. Whether this is to be seen as co-operation or coercion is something of an open question, and it may be noted that the system originates solely from one influential client federation. The contract form related to it, but not required by the system, is the product solely of an association of architects in private practice working in conjunction with the BPF. It can hardly be regarded as biased towards contractors, but rather as favouring the client more than does even the GC/Works/1 contract (see Ch. 15).

As indicated in Chapter 4, there is a basic version of the ACA agreement, upon which the version discussed here is based. That version allows for the alternatives of the contractor either performing or not performing the design, while the present version always requires him to design. It is otherwise almost identical in substance and detail at the points taken in this chapter, which may therefore be taken as applying to it also, subject mainly to ignoring references to the BPF system and manual.

Some features of the system

The BPF system is intended for 'medium-sized projects' in the private sector, but also to be capable of adaptation in the structures described, so that small or large projects can be covered. It is intended to be flexible enough for schemes which are complex or simple and need design and construction teams to match. It is a central part of the system that pre-contract design rests solely with the client's consultants (the term 'employer' is not used), and that post-contract design is given solely to the contractor and his sub-contractors.

Organisational responsibilities

Another key feature is the introduction of a client's representative, who acts as project manager and also takes many of the responsibilities of the architect under the JCT standard forms. This is more akin to the position under the GC/Works/1 and ICE contracts and is substantially distinct from the role of the employer under the JCT with-design form, where he is acting simply as a party to the contract with his own interests to look after. Even the employer's agent under that form, if appointed, exercises the same functions so far as delegated by the employer. The present client's representative speaks for the client (to whom he must refer back over many issues) to both consultants and contractor, but he also acts between the parties under the agreement in such matters as resolving documentary ambiguity and issuing certificates. This duality has the potential for more disputes than the JCT forms, although such evidence as is available suggests that this is not significantly so in practice. As with the GC/Works/1 contract, this is possibly due to the strong position of the typical client.

Within the system, however, the client's representative is not defined as a designer. A design leader is in charge of the team performing pre-contract design and also sanctioning the contractor's post-contract design. The design leader and his team are not mentioned in the agreement and must be dealt with through the client's representative. The manual recognises that the boundary between the duties of the client's representative and design leader may shift somewhat, but not that either may be eliminated. It is also possible that the design leader may change between the pre-tender stages. There is also provision for separate optional cost consultants to both the client's representative and the design leader.

This organisation structure means that the client deals primarily with one person, his representative, in respect of design and construction, but with more persons in all than is traditional. It needs very clear definition of the division of duties, both design and cost, between the various levels of authority and the various pre-tender stages. It is intended that pre-tender design responsibility should rest with the design leader, who must therefore clarify further responsibilities in his team carefully, especially as cost responsibility goes with design here. In addition, consultants' agreements are so drawn up as to make them more liable to penalties if there are faults in what they do.

It is not only the client who needs reassurance here, as both the client's representative and the design leader are *directly* responsible to him, but also the contractor. He alone is responsible for taking over and developing the design under the agreement and his responsibilities are spelt out quite fully, as discussed hereafter. As is also noted, the agreement is unbalanced by giving no indication that the client has any responsibility for his consultants' design work, and perhaps none is intended. This is neither reasonable nor realistic, and the contractor should seek 'clarification' over issues of the sorts highlighted. This is particularly important as the contractor has no opportunity to influence pre-tender design, or to put forward design proposals with his tender, so leading to some measure of acceptance and agreement. Areas discussed hereafter include:

(a) How the contractor's design is sanctioned.
(b) Whether any adjustment of the contract sum may arise from design development.
(c) Who carries liability for design defects, and what this liability is.
(d) How the contractor and his sub-contractors are related in these matters.

While complexity is sometimes needed to secure precision and control, it may be suspected that there is some duplication of effort in the present system in design and cost control, leading to additional expense and new uncertainties in place of the old, which the BPF is rightly against.

Procedural stages

The manual isolates five stages, which may be compared with the RIBA stages in Chapter 5. Four of these are pre-tender and so before the agreement comes into action.

Stage 1 – Concepts
This is the first evolution, including the general feasibility and outline cost plan. The client's representative may be appointed at this stage.

Stage 2 – Preparation of brief
This starts with the appointment of the client's representative and design leader.

The client's representative is in control of the total process and develops the programme and brief, with advice from the design leader. This should allow outline planning permission to be obtained, if required at this stage. The client is to decide, with advice from the client's representative, which parts of design are to be performed by the design leader and which by the contractor. This means how far the design is to have progressed when tenders are invited, after which responsibility passes to the contractor. Design and construction costs are identified within the total cost plan.

Stage 3 – Design development
The design leader, co-ordinating any further consultants, becomes responsible for 'expand(ing) the brief into a description of the required building in the form of drawings and specifications'. There are provisions for lump sum fees to consultants and for 'rewarding' them for keeping to time for design and budget for construction. There are several aspects here which consultants should weigh carefully, but which cannot be pursued. Development should take account of the client's requirements and allow him to see that they are being met, while also allowing detailed planning permission to be obtained.

Stage 4 – Tendering documentation and tendering
The prior information is to be refined for tendering and to be 'comprehensive, accurate, clear and unambiguous'. Among the factors involved are:

(a) A presupposition of firm design. The system does not allow incomplete design, which the consultants will polish up, or a tender sum which is transformed into a final sum by remeasurement or an equivalent process.
(b) No provisional sums for subsidiary parts are allowed: when there is doubt an assumption must be made and modified later under the variation provisions.
(c) Whatever has not been defined in full detail, must be specified sufficiently for the contractor to price what is required when tendering, and complete the design thereafter without price adjustment.
(d) The absence of the contractor's proposals accompanying the tender, although alternative proposals may be included. A schedule of required information with times forms part of the contract, but design follows.
(e) A schedule of activities replaces the bills of quantities as the standard analysis of the tender, although bills can be accommodated. This schedule is intended to relate to the sequence of work, rather than to cost elements as with the JCT contract sum analysis.
(f) Completion of the design by the contractor is subject to sanction by the client's representative, but without reduction of the contractor's responsibility.

Stage 5 – Construction
The contractor is responsible for construction and outstanding design, including obtaining statutory approvals. The client's representative remains as overall co-

ordinator, with the design leader and his team to examine the contractor's design and with a supervisor to monitor what is constructed. The schedule of activities is used for programme and financial control. In addition to arbitration after completion, there is provision for a separate adjudicator to deal with disputes during progress by a 'simple procedure'. The adjudicator's decisions may be taken to arbitration.

Documents under the agreement

Contract documents

The agreement in article C gives several other documents as 'the Contract Documents which form part of this Agreement':

(a) Obligatory:
 (i) contract drawings
 (ii) time schedule
(b) Optional:
 (i) schedule of activities
 (ii) specification
 (iii) bills of quantities

It is intended under the system as described, that (b)(i) and (ii) are normal, but with the option of (b)(iii) coming in either as additional or in place of (b)(ii). Clause 1.3 gives the agreement as prevailing over the other documents, unless the clause is amplified by specially inserted provisions. This needs to be done with care, as has been emphasised earlier in this book. Nothing is said about the relationship of the other documents to one another as a standing matter, except the points noted in the next paragraph. The time schedule is intended to be quite detailed, imposing strict obligations upon the contractor and so, indirectly, on the client's representative and consultants.

Under the JCT with-quantities forms, the contract bills contain the quality of the works, leaving the drawings by implication to show the general arrangements. Errors etc in the bills are then to be corrected and the contract sum adjusted. In the present agreement the place of the drawings is clear, but the relationship of specification and the bills needs clarifying, if both are used, preferably by allocating quality to the former and quantity to the latter. The second alternative clause 1.4 is similar to the JCT provision, by allowing correction of errors etc in the bills, but would also accommodate the specification used in this way. The inference is that the contract bills have been issued from the client's organisation and have not been prepared by the contractor. There are no rules about conformity to any particular method of measurement, and the bills need to give these rules, so reducing the area of doubt over whether mistakes have occurred. The first alternative clause 1.4 puts the schedule of activities in a similar position to that of bills as an analysis of the contract sum, but allows no financial adjustment for error in it, as it is prepared by the contractor. The other contract documents prevail if there is any clash. They all emanate from the client as available for tendering.

Clauses 1.5 and 1.6 allow the client's representative to issue instructions, variously resolving ambiguities and discrepancies in the contract documents and resolving infringements of statutory requirements. Adjusted payments follow on these instructions, except in two instances. One instance is ambiguity or discrepancy in or due to the schedule of activities or mistake in the contract bills, whichever applies under clause 1.4.

The other instance is the exception at the end of clause 1.5 over the contractor finding or foreseeing an ambiguity or discrepancy within the contract documents. This is qualified in two ways, both of which should alert the contractor. There is what the contractor could have discovered 'at the date of' the agreement, and so more than just while he is tendering. If amendments or qualifications to tendering documents are introduced before the tender is accepted or extra documents are added, and the manual allows these possibilities, the contractor should check their effect carefully.

But there is also what could 'reasonably' be found 'by a contractor exercising the standard of skill, care and diligence described in Clause 2.1'. That clause is considered further under 'Contractor's design responsibilities' below, but refers to the contractor's competence. Even with mistakes in quantities removed from its ambit, this is quite a drastic provision, requiring the contractor to check and correlate documents closely. It compares with the JCT with-quantities form which puts no retrospective obligation on the contractor of finding before commitment, or of notifying if he does, while instructions are issued and payment is always adjusted. The JCT with-design form goes a little further by allowing the contractor to choose between the discrepant options, although a change may still be instructed. Under the present agreement, what the contractor might 'reasonably' find in the usually hectic tendering period, when he is reading the documents primarily from a different angle, is perhaps to be interpreted in a very limited way. Restriction to cases in which he has attempted to manipulate meanings or has suppressed the obvious comment may be expected. Anything extreme might run foul of the unfair terms legislation. Even less satisfactory is the wording 'reasonably have been found or foreseen'; in that the clause is dealing solely with contract documents, incipient ambiguities etc appear meaningless.

Post-contract documents

While various documents like instructions come from the client's representative, the contractor is responsible for producing any remaining design and specification. Clause 2.2 gives three classes:

(a) Reasonably necessary to explain (to whom is not stated) and amplify the contract documents.

(b) Reasonably necessary for the contractor to perform the works and comply with instructions.
(c) Stated in the contract documents.

Leaving aside the subtlety of these distinctions and the overlaps possible, (c) allows for such categories as operating data and as-built drawings.

Clauses 2.3 to 2.5 deal with the procedures for the client's representative sanctioning what the contractor produces. The contractor is already under a programme obligation by the time schedule and the client's representative has to respond within the period set out in the clause. He may return documents with or without comment. If he makes comments, 'the Contractor shall immediately take account of such comments' and resubmit. What account he is to take is not specified, it being remembered that he is essentially developing the design and specification as allowed in his tender, without adjustment of the contract sum, but with responsibility for results. There is room for adjudication and even arbitration here. At any rate, when he receives documents back without comment, they are to be used for the works.

The client's representative's side of this procedure is related to the next heading, as is clause 2.6. Clause 3.5 out of that next group of clauses may be noted here: it allows the client's representative to require samples of materials or work to establish quality, with the cost either included in the contract sum or reimbursed as an addition, according to the contract documents. Further, the contractor may seek acceptance of samples in lieu of drawings etc that he would otherwise produce: here he bears the cost without extra payment.

Contractor's design responsibilities

This subject is the main thrust of clause 3 needing discussion, along with related clauses. Article B records that the contractor has offered 'to execute and complete the Works', while article F refers on to the provision by him of documents as discussed.

Clause 1.1 repeats the obligation to perform and clause 1.2 gives the 'skill, care and diligence' criterion mentioned above in relation to discrepancies. It is that of 'a properly qualified and competent contractor' dealing with this type of works. In terms of the related discussion in Chapter 5, this is in itself 'general skill and care' obligation with the addition of 'diligence', rather than a 'strict care' obligation, for all that it is preceded by 'all the'. So far as construction goes, the contract documents should prescribe precise standards to be achieved, so giving a duty of strict care in those respects over materials and workmanship within the general duty. This is similar to the duty under the various JCT standard forms as distinct from the with-design form.

The client's pre-contract documents are intended to develop the design well beyond the schematic stage, so that the contractor's responsibilities are essentially detailing, based upon his pre-contract work but not necessarily made available (even if existing) until the post-contract stage. This places the contractor's work across the whole project in a similar category to that reserved for parts of a project only under the provisions for performance specified work. These are considered in Chapter 17 and their detailed procedures may be compared with the requirements here. Broadly in this other case, the portions of work are usually intended to be self-contained elements of work and so development of their design and specification can be undertaken by the contractor in relative isolation from that of most of the rest of the works, so far as functional efficiency is concerned. The co-ordination of design will lie with the architect. In the present case, the contractor will often need to take more account of several elements together when performing detailing to give a satisfactory solution, because of the interaction of functions and his wider responsibility for design. The remarks following should be read in the light of these comments.

Details

In the present agreement a part of execution is design and specification, taking on from the position at tender, that is, from what the design leader and his team have done. It thus falls within the 'all the skill, care and diligence' orbit but, as indicated in Chapter 5 over unqualified design and construction by the contractor, this for the post-contract design could mean strict care, or achieving 'fitness for purpose'.

This is in fact made an express warranty of the contractor in clause 3.1 in respect of 'those parts of the Works to be designed by the contractor'. Thus design of windows would be required to lead to stability, weathertightness, resistance to rotting or corrosion, proper opening parts, lights and thermal characteristics specified and so on. It would not extend to the wider contribution made by the windows, such as internal comfort conditions and elevational appearance, unless facets of these matters had been so specified. This adds point, for both parties, to the manual's stress in stage 4 of the system on precise delineation of what the contractor is to design. It is not only that the contractor has to be sure what he is to design and so to install, but boundaries over responsibilities have to bet set.

Performance

The parallel warranty in clause 3.1 is that 'the Works will comply with any performance specification or requirement'. This is far more difficult to interpret, when the earlier design stages have in no way involved the contractor. It is the intention of the system that as much detailing as possible be left to the contractor, but it appears to be the normal intention that the major decisions, and perhaps others, should be taken by the design leader. In fact, the manual gives the illustration of the contractor designing reinforcement within a concrete beam, already designed in profile by a consultant (not necessarily the best way to optimise the result). The system is not dependent upon prior design going so far, but the further it does go, the less room there is for the contractor to make adjustments to achieve a performance specification for 'the Works' in some global way, rather than a part. Even more care (and skill?) is needed here to rest liability fairly (in all senses) on the contractor.

Both warranties are in the context of a statement that the contractor is to be responsible for all mistakes etc in the post-contract documents which he produces. This is fine in that he will have to bear the cost of dimensional rectification during construction, and should remain liable for inadequacies in the finished works which are due to his default. This is taken up in the disclaimer in clause 3.4 that 'no comment or advice from the Client's Representative' on information provided by the contractor is to 'relieve the Contractor from his responsibility'. This is making a point on which there is silence in the JCT with-design form although, as pointed out several times in Part 2 of this book, there is a strong presumption there for the same position.

Division of responsibility

What is not so fine, is the lack of any statement over the design responsibilities of the client's representative or his consultants, in the face of the various allocations of responsibility to the contractor. While the contractor is not expressly required to check earlier design and assume responsibility for it, he is perhaps being forced to go so far, if he is to produce design solutions and meet performance specifications applying to the works quite widely. The discussion under the next heading about foundations indicates the sort of situation which may arise. Even more acutely, he has in theory to do so before putting forward his tender, because his price will not be adjusted to take up unforeseen design provisions. More probably, he will make assumptions which he hopes cover his risk. This is a return to the old theme in the industry that the contractor first prices the job, and then he prices the architect!

It would be fairer, and so lead to lower tenders and perhaps lower final accounts, if the client assumed express responsibility in the present system for his consultants' design work, in all cases in which the contractor could develop it without needing to go behind and check aspects before proceeding with his own work. The question may not be critical when the contractor is merely specifying and providing a component or material, but is likely to be when performance of the works or even a part is really involved. In the absence of a standard clause, the contractor is advised to seek a special insertion in clause 1.3 and to state with his tender any aspects of his design that are based upon the qualification. He will still need to ensure that he has adequate tendering time to devise his solutions, in view of the fixity of his tender.

The contractor is strongly advised always to obtain indemnity insurance in respect of his design, and clause 6.6 obliges him to do this to a limit of amount given in the contract documents. This is against negligence on his part, or that of his sub-contractor etc and so is in respect of infringement of his duty of general care only. While the contractor's insurance covers those for whom he is responsible, they in turn are advised to cover themselves against the possibility of subrogation. Design insurance is discussed more broadly in Chapter 5.

Although the contractor is not required to produce his design and other documents until he has obtained the contract, he has to present them to the client's representative for comment under clauses 2.2 to 2.5 before he proceeds with related work. The uncertain element in this has already been mentioned. If therefore any major design issue is at stake, the contractor should seek the comments of the client's representative when the tender is being assessed. This need not mean putting forward full design, but data equivalent to contractor's proposals under a full design and build arrangement. This avoids too drastic a divergence later.

Groundworks

The optional clause 2.6 about 'adverse ground conditions or artificial obstructions' is similar in effect to provisions in the GC/Works/1 and ICE contracts. It requires the contractor to notify the client's representative of such problems and how he proposes to deal with them. The client's representative then instructs the contractor, not necessarily taking up the contractor's proposals in so doing. The contractor is reimbursed for complying with this, as with most instructions, unless he 'could reasonably have . . . foreseen' the situation at the date of the agreement (effectively when tendering), or unless the correction of bills of quantities mistakes covers reimbursement.

As with the other two contracts, there are two aspects underlying this provision. One is the question of working in adverse or artificial conditions, where the mode of excavation, disposal, supporting faces and dealing with water are common problems to resolve. The wording of the clause is similar to that in regular non-design and build contracts when they consider the question at all (the JCT group do not). Such wording in those cases allows for the extra costs relating to excavations to be reimbursed to the contractor unless he should reasonably have foreseen the conditions. Even then, provisions of standard methods of measurement supporting bills of quantities provide that some of these conditions always rank for additional payment without recourse to the not 'foreseen' proviso.

The other aspect of the provision is that of what is often additional constructional work (although it may be less) in foundations etc arising from the ground conditions. Under non-design and build contracts, the extra costs of this fall upon the client as the design implications are resolved by his consultants and the results are valued accordingly. The wording in the present clause is wide enough for the contractor to be reimbursed for additional work of this type also when this could not reasonably have been foreseen. As against quantities contracts, however, he apparently should allow here for such extra work which *could* be foreseen when tendering.

The extent of this position is not altogether clear: there may be, for instance, inadequately designed foundations shown on the client's drawings made available to tenderers, as against what is foreseeably needed in relation to low bearing capacity subsoil. Much will depend on just what the contractor is asked to do by way of design, whether it is limited, for instance, to detailing reinforcement within given profiles (as the BPF manual instances for columns) or whether it stretches to redesign of foundation sizes or even types. It is the question of how far the client underwrites prior design as already considered, so that the contractor can take it as read. Prudently he should state in his tender what he is assuming, particularly as his design is not to be produced until after entering into the contract.

The clause is optional and should only be included when really needed. Alternatively, its effect might be modified to limit the contractor's risk in some specific way in the light of the comments just made.

Design-related provisions

Sub-contractors

Clause 9.2 contains the usual requirement for consent to any sub-letting by the contractor. The manual emphasises the value of the contractor declaring names where possible with his tender. There are no provisions for nomination, but clause 9.3 allows for the naming of one or more persons in the contract documents to perform 'certain work' or provide 'certain goods and/or materials'. If so, the contractor prices the work when tendering and has the choice, if any, over whom he selects. It is indicated in the manual that such persons may have been consulted during pre-tender design and that a restricted list is desirable. Nevertheless the manual suggests, as the clause does not, that the contractor might be allowed to propose other names. Clauses 9.4, 9.5 and the first part of 9.6 are not used in the BPF edition, as they relate to the sub-contracting of work covered by provisional sums, which the system does not use. There is therefore no power available to the client's representative to impose a sub-contractor on the contractor after the date of the agreement.

Two lapse situations are covered in clauses 9.5 and 9.6: respectively, inability of the contractor to enter into a sub-contract with *any* named person, and termination of a sub-contract. In the former situation, this must be for a reason beyond the control of the contractor, whereas in the latter no reason is required. In both situations the contractor has to select another person, not necessarily from the earlier list, for approval of the client's representative. The clauses run against the pattern of legal decisions over *nominated* persons, by providing that the contractor receives no adjustment of the contract sum or other extra payment, and no adjustment of the time schedule. It would also be the case that he would have to bear the effects of any redesign, if proprietary components etc were involved.

These various clauses say nothing about design, whereas clause 9.8 makes the contractor 'fully responsible' for any sub-contract design, in terms equivalent to those already discussed over his design generally. This extends to meeting performance specifications and to the contractor's co-ordination of design, so that he needs to take all aspects into account. This is quite different from the position with a nominated sub-contractor's design, brought out in Chapter 14. It is more likely that obtaining in full design and build, subject to the preceding design carried out by the design leader and to the possibility, in the case of a named sub-contractor, that he will have been involved in that design. The optional sub-contract (it is not mentioned in the agreement) contains clauses equivalent to those in the agreement and the contractor should insist that sub-contractors enter into it, for his own protection. He will, however, remain liable if the sub-contractor goes out of existence. (In passing, if the contractor is entering into the ordinary ACA agreement using the alternative clauses under which he does not perform design, he should ensure that clause 9.8 is deleted, as this would make him responsible for design about which he knows nothing.)

Three situations arise:

(a) If the sub-contractor is not named, the contractor controls the basis of any design.
(b) If he is named, the position may be the same.
(c) If, though, the sub-contractor has been advising the design leader or other consultants prior to the tender, the contractor is dependent upon the sub-contractor for the design basis.

In this last instance the normal relationship is inverted and the contractor should check closely that the sub-contractor's proposals and the tendering specification tie up, in case anything has been changed since the advice was given. Even though the consultants and the sub-contractor have had discussions, the contractor will still be liable for any design which is prepared by the sub-contractor, and which strictly is put forward later than the tender. Only design embodied in the consultants' work issued for tendering may not be the contractor's responsibility, and this is subject to the cautions already expressed.

Actions of client's representative

The client's representative has a right of access by clause 4.1 to the works and elsewhere, as is normal practice. This specifically includes access to 'places where . . . design is being prepared', although this would reasonably be inferred.

The authority of the client's representative to issue instructions affecting the design runs up to taking-over of the works under clause 8.1, although its operation may be modified by some of the provisions for valuing the results in clause 17, which give a special procedure for agreeing the adjustment in advance, subject to some flexibility (not pursued further here). Paragraph (d) of clause 8.1 restricts the instructions to 'the alteration or modification of the design . . . of the Works as described in the Contract Documents'. This might be read as 'the Works which are described in', so that the client's representative could modify directly both pre- and post-contract design. This would lead to confusion of responsibilities, and the reasonable interpretation is to read 'the design . . . as described in'. This allows the client's representative to instruct over what is in the original contract only, so producing the equivalent to a modification of the employer's requirements under design and build. The contractor then has to respond by modifying his design to suit, and he alone may do this. This accords with the term in clause 2.2 over contractor's documents 'reasonably necessary . . . to comply with any instruction'.

The final certificate issued by the client's representative under clause 19.2 is, by clause 19.3, solely a statement of the final balance due. Clause 19.5 states that it does not 'relieve the Contractor from any liability'. The contractor should consider this carefully in the light of the comments about design made earlier, in addition to his more regular liabilities over work and materials. No doubt consultants will consider equally carefully their special liabilities to the client in their agreement under the system and over comments on the contractor's design that they make through the client's representative in the present agreement.

Financial provisions

The matters covered so far indicate that this contract has a lump sum basis and that the contractor carries the risks associated with design development post-contractually. The broad sweep of client's representative's instructions is similar to the architect's in the JCT contract. There are differences, such as a more demanding provision about acceleration, but these do not fall within the present scope of discussion.

Significantly, there is a mandatory procedure of prior quotations for the effect of instructions, covering the normal variation effect, any extension of time, any resultant loss and expense and any revisions to the schedule of activities. The client's representative may choose to dispense with the procedure, when normal valuation after the instruction follows. Otherwise there is a tight timetable for provision and agreement of estimates before work is to proceed. The client's representative has three options when there is not agreement: to instruct the contractor to proceed with valuation following the work, to withdraw the instruction, or to refer the estimates to the adjudicator for a decision. The adjudicator's decision is binding until after completion; it may be accepted by both parties or lead on to arbitration.

The contractor can effectively negate the prior agreement approach by simply doing nothing about a quotation or by putting forward an unacceptably high quotation and declining to reduce it. The former may be a reasonable course of action, or rather inaction, when the work or its disturbance features cannot reasonably be forecast. The latter is perhaps a more awkward situation, as the quantity surveyor then starts his valuation knowing that dissent is almost certain. However, if the contractor does nothing it is provided that he shall receive no payment for the results of the instruction until the final certificate, and then payment is without interest. Presumably his view on this will be tempered by whether the instruction leads to an extra or a saving!

This procedure for quotations is additional to anything in the mainstream JCT conditions, although there is a procedure in the supplementary provisions considered in Chapter 14. The GC/Works/1 and ICE contracts also adopt this approach (see Chs 15 and 16).

Chapter 14

Supplementary provisions of the JCT contract

General provisions

Named sub-contractor provisions

Financial provisions

The 'Supplementary Provisions (issued February 1988)', to give them their full title, are introduced in Chapter 4. They are placed after the main conditions of the JCT with contractor's design contract, having been introduced primarily to enable parts of the BPF system approach to be accommodated by the JCT contract. The BPF is a constituent member of the JCT. The provisions make no reference to the BPF system and they may be used as they stand without importing that system into a project.

Even when the BPF system is in use, these provisions allow for an approach somewhat different from the system outlined in Chapter 13. In relation to the BPF system they *amend* its operation, largely at its stage 4 'Tendering documentation and tendering', while they *expand* the JCT contract, largely at BPF stage 5 'Construction'. As the BPF manual has no contractual significance, it is necessary for the employer's requirements to spell out the parts, if any, of its approach and procedures which are to be incorporated in a contract including the present provisions. In particular, contractor's proposals still feature as a regular part of the present contract system, and have the same status as in an unsupplemented contract, while they are absent from the BPF system where the contractor apparently does not reveal his design detail until the post-contract stage.

While the BPF system was in mind in the drafting of the present provisions, they are in no way dependent on the system for completeness and may be used by any employer who seeks the extra facility and control which they give over such elements as dispute resolution, prior inspection of drawings etc, site control, selection of sub-contractors, pricing of change instructions (with an option of bills of quantities within the employer's requirements) and loss and expense.

The provisions read as complete clauses essentially intelligible on their own, rather than as a series of fragmentary verbal amendments to clauses, as happens with the designed portion supplement considered in Chapter 12. The several clauses operate largely independently of one another as modifications to, or amplifications of, the basic conditions. They may therefore be chosen separately without the need to use the entire set, subject to dealing with any minor interconnections, if this suits the particular scheme or employer. It is important that the employer's requirements give the additional information to back up various clauses.

The supplement is mentioned in article 1 of the JCT contract as modifying the basic conditions, but only when the entry in appendix 1 is suitably dealt with by deletion of one alternative out of 'to apply/not to apply'. This is an inconspicuous place for such an important item and it should not be overlooked. If the deletion is not made there will be uncertainty, which hopefully will be resolved by reference to the intention of the parties in the other documents, such as is shown by the additional backing information.

General provisions

Clause S1 – Adjudication

A dispute resolution pattern is introduced to be used instead of or as a prelude to arbitration under clause 39 over specified 'Adjudication Matters' of some significance:

(a) Adjustment or alteration of the contract sum.
(b) Execution of the works in accordance with the contract.
(c) Empowerment of an instruction under the contract.
(d) Unreasonable delay in giving consents, statements and agreements.
(e) Differences over aspects of obligations and restrictions, partial possession or occupation of some part of the site by the employer, design sub-letting and extension of time.

Even if arbitration may follow, the adjudication procedure must be used first and a decision stands as 'an Adjudicated Provision' until any arbitration overtakes it. It is available only for disputes arising before practical completion or its equivalent, but this does not prevent it actually taking place after that time.

The procedure sets an adjudication in motion by reference to a contractually named or subsequently appointed adjudicator, much as for arbitration. By clause S1.3.3, his decision is 'final and binding' unless one party informs the other under the next clause that he wishes the matter to be referred to arbitration. Subject to arbitration, the decision of the adjudicator prevails over 'any other provision of this Contract' if there is a conflict, giving it considerable force in contract interpretation.

Clause S2 – Submission of drawings etc to employer

Under clause S2.1, the contractor is to submit drawings and other documents which he prepares to the employer before use, which advances from the silence of the main conditions, but only so far as this is provided for in the employer's requirements. The employer has whatever 'rights . . . in regard to his comments' may be stipulated in the requirements.

This allows the employer plenty of discretion over requiring information. However when he receives it, he should still avoid giving explicit approval on the 'appears to meet' basis discussed in several places in this book (see eg under 'Design responsibility' in Ch. 5), so that responsibility remains with the contractor. Clause S2.2 acts to keep responsibility generally with the contractor, but may be neutralised by 'comments specifically so [stating]' otherwise.

Clause S3 – Site manager

If the employer's requirements say so, clause 10 is replaced by this more demanding provision of a designated full-time manager approved by the employer. The manager is not to be replaced without the employer's consent, he is to attend the employer's meetings (along with other construction personnel, if required) and he is to keep any records stipulated in the employer's requirements.

Named sub-contractor provisions

Clause S4 – Persons named as sub-contractors in employer's requirements

This clause departs from the standard philosophy of the JCT with contractor's design form that the contractor is allowed complete flexibility over arriving at his proposals, once the employer's requirements have spelt out the nature of the end-product. Clause S4.1 allows 'Named Sub-contract Work' to be reserved in the employer's requirements for performance by a 'Named Sub-contractor'. No other stipulation is made about how this information is to be given, although a clear and complete statement of the scope and nature of work to be performed must be given, such as (in principle) 'all windows and external doors and these are to be in aluminium' or 'the structural frame and this is to be laminated timber'. This will bind the contractor's choice on that work and possibly in consequence on other related work, so that his scope for design solutions is limited.

It is generally best if this approach is limited to work which is substantial in extent, essential to the employer's needs and quite specialised. It is likely to be suitable when there has been detailed discussion with the sub-contractor, to enable the employer's other requirements to be framed around it, and when the sub-contractor is specially suited to provide a solution. The clause assumes that only one sub-contractor will be named, rather than that something like the 'list of three' arrangement included in the JCT standard form at clause 19.3 and discussed in Chapter 7 in relation to the with contractor's design form. It makes no reference to the price of the initial scheme.

Clauses S4.2.1 and S4.2.2 provide that the contractor is to enter into a sub-contract soon after entering into the main contract. The employer, if notified by the contractor of a problem in doing this, is to act by a change instruction (and so affect the final amount payable) to 'remove the reason for the inability' of the contractor to enter into the sub-contract. Alternatively, he may omit the work as named sub-contract work by a change and instruct its execution by one of two routes given in clause S4.3. These are by instructing the contractor to select another person to perform the work or by having the work performed by a direct contractor related to the present contract by clause 29. The employer may not 'name' another sub-contractor in his instruction in the first of these instances.

Clause S4.4 deals with determination when once the sub-contract is in being. The contractor may determine on the sub-contractor's 'act or omission' but only with the employer's consent. Nothing specific is stated about the sub-contractor determining and there are no published sub-contract conditions. However a determination occurs, 'the Contractor shall himself complete' any outstanding work. It is suggested that 'himself' is emphatic that there may not be another 'naming', rather than that the contractor may not sub-let the work with approval.

The financial settlement when there is determination is for the result to be treated as a change, unless the contractor is at fault by determining without consent or 'by [his] act or omission' so that the sub-contractor determines. The contractor is to recover whatever he can from the named sub-contractor 'by reasonable diligence', that is by all means short of legal action, and account for these amounts gross to the employer. Beyond this he is not liable to the employer. Clause S4.4.4 is given as a result and requires the contractor to include a term in the sub-contract that the sub-contractor 'undertakes not to contend [in proceedings etc] . . . that the Contractor has suffered no loss [etc]'. Otherwise the sub-contractor could escape at least some liability on the basis that the employer had relieved the contractor, who would not have a loss to press against the sub-contractor. This is a rather unusual stipulation also used in the JCT management contract, discussed in the present author's *Building Contracts: A Practical Guide*.

This pattern has a number of similarities to that for named sub-contractors in the JCT intermediate form of contract (also discussed in *Building Contracts: A Practical Guide*), but lacks any statement about a prior tender provided to the employer for the sub-contract work. Several issues stated or implied in the present provisions may be set alongside each other to highlight the position resulting:

(a) The sub-contract work must be clearly and completely defined in the employer's requirements, if the contractor is to take it into account in his tender over both design and price, but it cannot necessarily be firmly quantified to give a fixed lump sum price. This will depend on how much latitude is afforded by the employer's requirements for layout variation on the contractor's part.
(b) There is no statement that the employer has obtained a tender on any basis ahead of the main tender for the work of the proposed named sub-contractor.
(c) Only one sub-contractor is to be named rather than a group, so following the

JCT intermediate form named pattern rather than the JCT standard form list of three pattern, so that there is no room for any measure of sub-contract competition after the main contract is in being. Some form of necessary negotiation is implied by the possibility that there may be an inability to enter into the sub-contract as required.

(d) The contractor is to enter into what is effectively a domestic sub-contract on terms undefined by the main contract and this is to be binding upon him in all respects, including a price adjustable only where the main contract terms provide over such matters as changes.

(e) There is to be a change instruction by the employer over an 'inability' on the contractor's part to enter into a sub-contract. Presumably in most cases this will either mean a direct price increase or one of the other arrangements allowed, which is likely to be at an extra cost to the employer.

The pattern is left apparently incomplete by elements (b) and (e), especially by comparison with the JCT intermediate form. Unless the employer knows the original basis of the named sub-contractor's tender, and so what the 'inability' to formalise is about, he has no starting point for his own action or, more, whether he is liable to take any action (and pay for it) at all. The lesson for the employer is clear, even if he has to frame matters rather flexibly in his requirements because the design is fluid.

Financial provisions

Clause S5 – Bills of quantities

Any bills of quantities included in the employer's requirements and so emanating from the employer form part of the contract. If the contract sum analysis provided by the contractor consists of quantities, the clause is not applicable. The method of measurement, which may be quite unorthodox, is to be stated. Errors are to be corrected as changes and the bills used for change valuation and purposes of formula adjustment. All of this follows the philosophy of the JCT standard form with quantities.

The practicability of using quantities in this way will depend on the state of design when tenders are invited. If there is little more to do than to complete detailing, as the straight BPF system envisages, this is quite possible. But the more open the design is and its supporting specification, the less reasonable will it be to employ quantities. The general problems associated with quantities as a contract sum analysis are discussed in Chapter 9.

Two variations on the quantities theme (if the expression be permitted) are possible. One is to use firm quantities for such parts of the scheme as are fixed by the employer and for these only to be governed by clause S5 and to appear as such in the analysis. This becomes a designed portion supplement (see Ch. 12) in reverse and is likely to be of very limited application. It is also likely to be of very limited value to either party.

The other possibility is to use approximate quantities for a significant part of the scheme, provided that the specification of work is known. If so, it will be necessary to amend the clauses given on the lines of what is in clause 13 of the JCT with approximate quantities form. For example, such quantities may be useful if the scheme is of system design, to be erected on foundations perhaps of traditional nature and with ground conditions which are difficult to determine in advance. The problem with approximate quantities in a design and build situation is that they may remove from the contractor the incentive to design economically, in that any over-design is paid for by virtue of the remeasurement. This problem was aired in Chapter 9 over valuing changes and also occurs under the ICE contract (see Ch. 16).

Clause S6 – Valuation of change instructions – direct loss and/or expense – submission of estimates by the contractor

This clause modifies the first two elements in its heading and adds the third, which is the critical point. It aims to introduce greater advance certainty into the whole change procedure, even more important than usual when there is no independent designer. The contractor is to submit estimates when a change instruction provokes a change valuation (as is almost inevitable), or leads to extension of time (not mentioned in the clause heading) or to loss and expense. Change estimates should be supported by sufficient outline design information, as is implicitly recognised by the closing sentence of clause S6. The outline procedure is similar to that under the GC/Works/1 contract, while the ICE contract has a simple equivalent. It is subject to tight time-scales and is:

(a) Estimates are always required under clause S6.2, unless the employer states otherwise or the contractor raises reasonable objection to giving estimates, presumably on the grounds that prior calculation is impracticable. This is quite likely over loss and expense, as the essence of such cost is that it relates to matters of disturbance, often by definition only to be ascertained after the event.
(b) The estimates under clause S6.3 (and presumably as agreed under clause S6.4) take the place of any other valuation or ascertainment and, it must be intended, any other fixing of a date by the employer.
(c) If agreed, then under clause S6.4 the estimates bind both parties, so precluding arbitration, unless some related rather than integral issue crops up.
(d) If the estimates are not agreed then, by clause S6.5 and on the employer's instruction or other initiative, they may be withdrawn and the instruction be complied with, or the instruction may be withdrawn (when the contractor is reimbursed for any extra and abortive design expense), or there may be a reference to adjudication.
(e) If the contractor fails to provide estimates, clause S6.6 provides for clauses 12, 25 or 26 to be revived unmodified, that is the estimate requirement lapses. However, there is a penalty introduced, as the contractor receives no

interim payment for additional amounts and no loss of interest or financing charges due to such delay.

The whole arrangement is thus subject to the routine problems associated with any 'agreement to agree', as happens with the next clause also. If the employer wishes to know his commitment before proceeding he has to give an instruction but, if needs be and the contractor is uncooperative, withdraw it quickly in the absence of an estimate. If the contractor has proceeded even more quickly with any work, the employer would appear bound to pay for it, despite the use of the term 'breach'.

Clause S7 – Direct loss and/or expense – submission of estimates by contractor

This clause deals with all loss and expense situations, except those due to change instructions covered in clause S6. Again, it modifies clause 26, but somewhat differently.

Firstly by clauses S7.2 and S7.3, it relates to estimates for loss and expense already incurred 'in the period immediately preceding' any application for payment, with a series of applications being made if the particular loss and expense run through more than one period.

Secondly by clause S7.4, the estimate accompanies the application for payment and so is not included in it. The employer has a tight timetable within which to act and may then give notice of one of the following:

(a) Acceptance of the estimate.
(b) Desire to negotiate and that, if there is then not an agreement, the matter is to be referred (upon his subsequent and unilateral choice) to adjudication or revert to being dealt with under clause 26 unmodified.
(c) Clause 26 unmodified is to be applied.

Thirdly by clause S7.5, whenever an accepted, agreed or adjudicated amount trickles down the cascade of (a) and (b), it is to be added to the contract sum. Payment to the contractor is therefore delayed by the time needed to settle and further until the next interim payment is due. Only if (c) is used directly, may an approximate amount be included on account by virtue of clause 3.

Lastly by clause S7.6, there is again a provision over the contractor being in breach by not submitting an estimate and being penalised to the extent of receiving no amount until final settlement.

Chapter 15

GC/Works/1, single-stage design and build contract

Contract basis and documents
Contractor's design responsibilities
Financial provisions

This chapter deals with the contract to which it relates only so far as its design-related implications are concerned. Even these are treated fairly briefly simply to set out differences of principle from those in the JCT with contractor's design contract covered in Part 2 of this book (referred to here as 'the JCT contract'). A rounded treatment of the mostly similar parent version of the present contract, the with-quantities edition, comparing it with the JCT standard form of contract on clause-by-clause basis, is given in the present author's *Building Contracts: A Practical Guide*. The present contract is introduced in Chapter 4.

The contract is published by Her Majesty's Stationery Office, having been produced by PSA Projects Ltd as consultants to the Department of the Environment. As such, it is a unilateral employer's production and not a consensus document and may be open to being construed *contra proferentem* in places as a result. This gives a distinction of emphasis which does not radically affect its interpretation for present purposes at least. It does, however, differ from the JCT contract quite radically over a number of matters, for instance extension of time and programme, the more active role of the client, the scope of instructions, loss and expense, determination and arbitration. These are not considered here except at points of contact with design, but they affect the overall operation of the contract extensively.

The result is a contract with distinctly firmer control of the project by the client than occurs under the JCT contract, which may be welcomed by some or interpreted by others as a bias towards the authority. As the document is to be marketed in a modified form as suitable for use by clients other than central government departments, this aspect should not be ignored by contractors and sub-contractors.

There is an introduction printed as part of the document which, it is stated, has no legal significance. Several explanatory points from this introduction may be mentioned initially which, with others, arise later in this chapter:

(a) The contract is one of a family, others being those with and without quantities. All are lump sum contracts, like the JCT contract.
(b) As its title indicates, it is for use when there is a single-stage competition and the client (termed 'the Authority') has precisely defined requirements (termed 'the Authority's Requirements'), while there is no need for control over the post-contract design (other than checking that detailing is not going astray) and there should be little likelihood of changes. The JCT practice notes make no definitive statement on these issues, whatever may be hoped for, while the JCT contract does not assume any particular stage of development when entering into the contract.
(c) It is stated that the contract is best for 'relatively simple buildings' and that for 'more complex buildings' there might be two stages of design, with the second stage 'after tender acceptance'. This appears to suggest before construction starts, and this is reinforced by the statement that *two* stages would need 'significant amendments' to the contract. Oddly, if the second stage was performed while site work was proceeding, the contract could reasonably be used without such amendment. The JCT contract is silent on the issues of complexity and stages, while clearly accommodating progressive design.
(d) Design responsibility is on a 'reasonable skill and care' basis. This follows the JCT contract, but there is a distinction in that there is also a requirement for the contractor to have design insurance. It is also hinted that fitness for purpose may be introduced, perhaps for limited categories of work, but no guidelines or provisions are given.

The cover gives the contract as 'for Building and Civil Engineering', in common with the rest of the series, which works quite well for the latter without the extra provisions in the ICE contract. However, as the present contract is in lump sum form but lacks the extra *design* procedures (such as over boreholes etc) of the ICE contract, it should be restricted to particularly simple civil engineering works.

Contract basis and documents

As stated, the contract is based upon a single-stage design performed before acceptance of the tender. This breaks even with the JCT contract in that 'The Design' is defined in condition 1(1) (there are conditions rather than clauses here) as consisting of the authority's requirements and contractor's proposals, and also all design documents for construction. The non-contractual introduction refers to these last as 'subsequent'. Procedural conditions discussed below are framed to allow for design development and detailing during progress.

The contract itself consists of documents equivalent to those in the JCT case and includes the requirements and proposals and also 'the Pricing Document', the term used for what the JCT contract calls the contract sum analysis. This appears to be similar in intent.

A significant distinction from the JCT approach occurs under condition 2. While the conditions here also prevail over the other contract documents, there is the difference that within these documents the authority's requirements are given as prevailing over the contractor's proposals if there is a discrepancy between them. This is therefore the opposite of the 'appears to meet' arrangement of the JCT contract, whereby the proposals implicitly and effectively prevail over the requirements if they are in conflict. It is in line, however, with the ICE contract taken in Chapter 16. Under the JCT pattern, the employer needs to ensure that any design development offered by the contractor in tendering fits what he wants, as that development is what will be in the contract and can be changed only through the normal mechanisms of change instructions. But more, that pattern allows the employer to be almost lax in expressing himself and to wait and see what is offered of interest to him by the one or more tenderers, so long as the proposals are sufficiently explicit and precise.

In the present contract it is the authority who needs (he is masculine singular in the conditions!) to be precise and explicit to ensure that he obtains something acceptable in the tender. On the other hand, the contractor must take care that what he offers by way of any *development* of the requirements does not achieve this by a *departure* from the requirements, that is he must not be imprudently innovative. The authority is entitled to ignore what is in the proposals and go back to what is in the requirements without price adjustment. It would even be possible for the authority deliberately, if unfairly, to accept without comment an offer based upon a cheaper design solution in some respects and thereafter to insist on the dearer approach outlined in the requirements. The opposite also applies, that the contractor may offer a dearer solution and then, when what he offers has been accepted (subject to competition), he may insist that he sticks by what is in the requirements without adjustment of price. Of course, the JCT provisions permit sharp practices in reverse instances.

It follows that the contractor should take care to see that he interprets the requirements properly in his proposals. If he considers that he has something more advantageous to offer, he should express this distinctly, by a query when tendering or by qualifying his tender. On balance, the contract philosophy appears to be aimed at projects which are fairly well developed in design when tenders are invited, rather than the reverse. Only so are the requirements likely to be firm enough to overrule the proposals if the need arises.

The position over other categories of discrepancies, that is those entirely within either the requirements or the proposals, is on the lines of the JCT contract and here partly differs from that under the ICE contract. Still under condition 2, the authority's project manager (the PM) is to resolve any discrepancy within the requirements or between them and statutory requirements and a change in the authority's requirements is held to follow. The implication is that the contractor when tendering has chosen the non-preferred alternative, or perhaps one out of several, and presumably this is the cheaper or cheapest. Otherwise he is likely to resist a priced variation. If there is a discrepancy within the contractor's proposals, the PM may choose to accept the contractor's proposed amendment of one of the discrepant options as given, without an adjustment of the contract sum resulting in either direction.

Contractor's design responsibilities

Insurance

Condition 8a requires the contractor to hold professional indemnity insurance over his design indemnity, something not required at all by the JCT contract. A discussion of the general question of design insurance is given in Chapter 5.

The period of insurance is to be from commencement of the contract (and so before work starts on site), through to certified completion and then for a further six years. This covers the period of contractual liability, as government contracts are not executed by deed but by exchange of letters, but does not cover any remaining period of tortious liability (see Ch. 10 on when liability expires). An authority of another type may wish to amend the period to twelve years. It is a difficult matter for the authority to police over such a long time, despite being given the power to ask the contractor for a certificate from the insurer throughout that the insurance is still in force. The contract is otherwise over in effective terms, subject to latent defects (which are what the insurance is really about), even though condition 39 on certifying work contains no provision for a final certificate on the lines of the JCT contract. There is no provision for the authority to insure if the contractor defaults at any time during the whole period and damages are of no value until the very events occur against which the insurance is needed.

The standard of insurance required is 'for design services on terms consistent with good professional practice'. This means insurance against a skill and care liability as what good professional practice regularly obtains, rather than against liability over fitness for purpose or strict care. This matches the design liability assumed by the contractor under condition 10. No detail is given of the policies, which may be 'existing or new' and 'as he sees fit', the latter term clearly relating to the former but apparently also to the policy details. There is regulation of the permitted insurer and the level of cover is to be at least the amount stated in the abstract of particulars. This amount is to be 'in respect of any one occurrence or series of occurrences'. This term relates to both the year of claim basis and the total protection afforded in the year, both of which are dealt with in Chapter 5. By comparison with the model Summary of Essential Insurance Requirements over all-risks insurance appended to the contract, these requirements are in quite broad terms.

There is no requirement about any insurance to be carried by any consultant of the contractor. This will be mandatory for any member of a recognised professional body and the authority has the facility to check who is proposed to perform design, as this is to be stated in the form of tender. Condition 62 on subletting makes no specific mention of designers, but requires provision of 'details of any subcontractor'. In any case, a consultant's insurance is not contractually available to the authority, although a route in tort may exist.

Design of the works

As a prelude to the mainstream matter of design, condition 9 on as-built drawings oddly starts with the end of the works and requires the contractor to supply 'full drawings' and other information to describe the works, with operational and maintenance details. This is a sufficiently variable area of interpretation for the information sought to be made explicit in the requirements.

Condition 10 on design of the works contains the main responsibilities of the contractor. By condition 10(1), he is to 'undertake and be responsible for the Design', so parallelling JCT clause 2.1. The design is defined in condition 1(1) as 'the complete design' and in accordance with the requirements and including both the pre-contract proposals and the post-contract design documents of condition 10a.

Design responsibility in condition 10(2) equates with what is in JCT clause 2.5.1 by requiring that 'The Contractor shall exercise, and warrants that he has exercised, all reasonable skill, care and diligence.' This again covers the proposals and the latter design documents. The other common elements are:

(a) Liability in respect of any defect or insufficiency in the design.
(b) Relation as a standard to 'an architect or other suitably qualified professional designer'.
(c) Comparison with the case of design prepared under a direct appointment by the authority for works to be executed by a works contractor.

There is no reference in the condition to any design contained in the authority's requirements. The inference of the definition in condition 1(1) is that they will contain none, but simply 'define' what is required. However, this is difficult to maintain in practice, as the process of definition often involves presenting some form of design which limits what is required and which must be taken over by the contractor. Further, condition 10(3) requires that the contractor shall ensure that the design meets the various statutory requirements, which include 'planning consents' some of which may have been given initially to the authority and then passed on to the contractor. Condition 10(4) states that the contractor is to ensure that 'all Things for incorporation' (a delightful piece of prose) conform to the requirements or otherwise to appropriate standards, where their mode of employment by the contractor may interact with his design.

All of this is common with the JCT contract, where again it is to be inferred. The contractor should take careful account of the precise implications in his project of any of these matters which affect it and, if necessary, qualify his proposals accordingly, so that uncertainties may be sorted before there is commitment. This is particularly urgent under this contract, as the requirements prevail over the proposals when the contract is in being.

According to the introduction bound in with the contract, the requirements are to define the site boundary; responsibility would clearly rest with the authority.

Condition 10a deals with the design documents to be provided by the contractor to the authority during progress. Condition 10a(1) refers to their supply in all cases to the PM before work is performed. It also stipulates that they shall 'demonstrate compliance on particular matters with the Authority's Requirements', although the term 'particular' has its usual vague connotation in such usage of 'anything in general'. 'Design Documents' are defined more widely in condition 1(1) as those 'prepared by the Contractor in the performance of the Contract for or in connection with the Design'. These are primarily those which the contractor prepares for his own use, but will normally cover the matters about which the authority wants to know. However, it is desirable for the requirements to set out and so particularise these matters, so that the contractor is not faced with a mass of unexpected drawings or schedules.

More critically in most situations, by condition 10a(2) the contractor may not commence work based on any design document until the PM has examined it and confirmed by endorsing the document either that he has no questions on it or that his questions have been satisfactorily answered. Condition 10a(5) states that nothing in the whole condition is to 'relieve the Contractor of any liability' over design documents. This confirms the inference to be drawn from the use of 'questions' in condition 10a(2): the contractor remains responsible in all ways for his design and the examination of drawings etc by the PM does not constitute approval or transfer of liability. This is in line with the position under the ICE contract, whereas the JCT contract leaves it to the architect's care in practical expression, except in its supplementary conditions. Condition 10a(6) stipulates that design documents are to be submitted within a period sufficient for examination and queries to take place adequately.

Discrepancies in or between design documents are to be resolved by the contractor to the PM's satisfaction under condition 10a(8) without any financial adjustment flowing. The existence of a procedure regulating provision of information to the PM is additional to anything in the JCT contract, except in its supplementary conditions, and the absence there of a mainstream arrangement has been noted adversely.

Condition 31 deals with 'Works quality' and requires the contractor to execute the works 'in accordance with the Design'. He is to provide samples and allow the PM to inspect the works, materials etc.

Because of the emphasis upon design, conditions 33 and 35 on programme and progress meetings both refer to the place of design and its monitoring by the PM. The programme is to give adequate detail for the several elements, so that the system of progress meetings, which features so prominently in this contract, can be used to keep check on what is going on.

Site conditions

Condition 7 on 'Conditions affecting Works' repeats precisely what is in the with-quantities version of this contract, subject to the omission from there of the inapplicable condition 7(2), referring to information in bills of quantities provided by the authority. The wording given in the present condition lends itself to wider interpretation.

Condition 7(1) deals explicitly with a range of site matters, which the JCT contract leaves either to be implied or to be covered in the employer's requirements. It lists conditions which the contractor is deemed 'to have satisfied himself as to', some of which extend far beyond the physical site. Several of the physical site matters relate to essentially superficial characteristics, such as contours, but the list includes 'the nature of the soil and material (whether natural or otherwise) to be excavated'. The reference in the with-quantities contract is to the problems of excavation and associated work and not to the design of what goes into the excavations. The term 'to be excavated' carried over into the present condition might be read as sustaining this here.

However, condition 7(2) requires the contractor to inform the PM of ground conditions of which he was not aware when tendering and of how he proposes to deal with them, and the PM is to certify under condition 7(3) if he agrees that they are 'Unforeseeable Ground Conditions'. If so, a variation arises, to be valued under the regular provisions. While under a regular quantities contract this might refer simply to such temporary expedients as the use of rock hammers or sheet piling, here the contractor is also responsible for design of the permanent works and it is inescapable that the wording 'measures which he proposes' carries the wider interpretation that more (or conceivably less) permanent work will be valued in the variation adjustment.

These provisions therefore remove the risk of the unforeseen in the ground from the contractor, while leaving with him conditions which were apparent from site investigations reports and so forth made available to him and also any matter which he could reasonably gather from wider local enquiries. They specifically exclude 'ground conditions . . . caused by weather'. They may be compared with the silence of the JCT contract and with the provisions of the BPF/ACA and ICE contracts. In the first, the employer's requirements need to be explicit to give any certainty. In the other two, there are provisions broadly comparable with those in the present contract, although the BPF/ACA clause is endorsed as optional.

Financial provisions

As with the JCT contract, the lump sum basis means that the contractor carries the responsibilities and risks associated with design development post-contractually. Clearly, the single-stage design and build approach to a simple project properly used should mean that the design is carried to a relatively advanced stage before the contract is awarded, so limiting the risks due to partly developed design or specification. The question of financial responsibility for discrepancies is covered under 'Contract basis and documents' above. The main run of financial provisions is contained in the contract section headed 'Instructions and Payment'.

Condition 40 deals with the PM's instructions on all matters in one place, the scope of which as listed is broadly in line with that in the JCT contract. There are one or two completely additional items which do not fall within the scope of this chapter. But also there is a final open-ended item of 'any other matter which the PM considers necessary or expedient'. This is standard practice in the GC/Works/1 family, but hopefully is not to be used too liberally in relation to aspects related to design, even indirectly. Its generality is overridden by the specific first item of the list 'a change in the Authority's Requirements'. This identifies what is to be instructed in parallel terms to what is in the JCT contract and the PM should follow the same principles over how he instructs, that is to change the requirements but not dictate the solution. This aspect is discussed in detail in Chapter 9.

There is a provision in condition 40(4) under which the contractor may be required to give a quotation and programme statement in advance relating to any variation instructed. This is similar to the provision in the JCT supplementary conditions and leads into valuation under the next three conditions.

Condition 41 deals with the principles for valuing all instructions, where the significant statement is that valuation is to 'include the cost (if any) of any disruption to or prolongation of both varied and unvaried work'. The stipulation is not peculiar to design matters or to this contract in its family, but differs from more orthodox stipulations in two ways. It requires disturbance costs to be calculated in advance and so not 'ascertained' as they are incurred: this may be difficult when disturbance is almost by definition irregular in its effects, although it should not be too difficult when the contractor is in control of the total design and the works are not complex. It also requires the costs to be allocated to each instruction, so assuming that several instructions do not have interactive effects. If necessary, the contractor will need to carry out overall calculations and then make some reasonable allocation between the instructions.

Condition 42 deals with the valuation of variations in particular. It provides two routes to deal with any one variation instruction: acceptance of a quotation as required by an instruction under condition 40 or valuation by the quantity surveyor, who is recognised under this contract. The former route is to be used unless either the PM does not require a quotation in the instruction, the contractor fails to provide it or agreement on its amount cannot be reached in advance. The JCT contract also has two routes for the final account as a whole, but these are quite different: a final account produced by the contractor or, on his default, one produced by the employer.

As with other such provisions in the contracts considered in this book, the contractor can frustrate their effect here by putting forward a quotation at an unacceptably high level and being unprepared to reduce it enough to allow agreement.

As a rider to the quotation approach, it is given that the quotation is to show separately 'the direct cost of complying' and 'the cost (if any) of any disruption . . . or prolongation'. Neither of these elements is necessarily to be related directly to the pricing document (the equivalent of the contract sum analysis). It is also to be a final commitment by the contractor, with no more payable. By inference, it is equally binding on the authority.

When the quantity surveyor values, he is to use 'such rates and prices' as are available and outline rules are given about changes in quantity and character. In addition, allowance is to be made for design costs, also 'on the basis of any rates'. Beyond this, 'fair rates and prices' are to be used, or daywork. Adjustment to the rates for any disturbance effects is to be made. The virtue of having a sufficiently fulsome pricing document is as clear here as with the contract sum analysis under the JCT contract. The wording given suggests that quantities may support at least part of the pricing document.

A procedure for notification by the quantity surveyor of his valuation and for any disagreement by the contractor is given. Again a time-scale emphasises speed in settlement, although if there is a disagreement the time-scale can impose nothing and breaks down.

The condition ends with a provision covering a variation instruction resulting in 'a saving in the cost' to the contractor. This is a rather odd expression in context and might be subject to equally odd interpretation, even more out of step with the pricing document. It is given that the quantity surveyor is to determine the amount, no alternative of a contractor's quotation being given.

Condition 43 provides more briefly for valuation of other instructions, that is those not covering a variation. It allows reimbursement of 'any expense' which is 'properly and directly' incurred or, again, the deduction of 'any saving in the cost'. The whole area of what is allowable as 'expense' and how this excludes other elements of what the JCT contract terms 'loss and expense' is highly significant, but beyond present discussion.

Leaving aside the detail, declaring amounts as soon as possible and agreeing them whenever possible before, during or soon after execution of the work instructed is a highly desirable aim. The active involvement of the quantity surveyor is made possible by his naming here. Stipulations about the role of the employer's agent in this way could well be made in the employer's requirements under the main provisions of the JCT contract, rather than in its supplement. There is no provision in the conditions for provisional sums; the introduction to the contract states that this is 'to reinforce the importance' of the authority giving precise requirements. The same paragraph recommends that exceptionally such sums may be given in the authority's requirements with supporting rules.

Chapter 16

ICE design and construct contract

Contract basis and documents

Contractor's design responsibilities

Financial provisions

Like that immediately preceding, this chapter deals with the contract concerned only over its design-related implications, and the comments in the opening paragraph of that chapter are equally applicable here, comparison again being made with the JCT contract in its contractor's design version.

It may be argued that a contract for civil engineering work has no place in review of design and *build* contract practice. However, there are a number of interesting elements in the way in which the contract treats design and construction procurement in its rather (but not radically) different context, which are worth reviewing here by way of comparison. It is also the case that building often contains substantial blocks of work of a civil engineering nature, such as heavy sub-structures and extensive external works, for which the provisions of the present contract may be worth studying, even if then adapted or ignored. These construction elements are often the most uncertain in design or extent, so that a tidy lump sum approach is at its most questionable in terms of risk distribution.

The contract is produced by consultant and contracting bodies involved in the civil engineering industry and is in its first edition; it is introduced in Chapter 4. It is derived directly from the sixth edition of the ICE conditions for civil engineering work (referred to here as 'the ICE standard contract') which deal in the traditional manner with design by the engineer (equivalent to the architect) and construction by the contractor, with complete remeasurement of the approximate quantities based works. Many clauses of the two forms are identical or differ only in drafting detail. All clauses of the other conditions are treated in the present author's *Building Contracts: A Practical Guide* by comparison with the JCT standard form of contract with quantities.

The contract is intended for use by a client (termed 'the Employer') in either the private or the public sector, although its use is predominantly in the latter. In practice, the long-established ICE contract gives quite firm control to the employer or the engineer, even though this is achieved by a measure of uncertainty in the wording of the conditions and the exercise by the engineer of much discretion and an authority which stems from the accepted practices of the industry, where engineers have operated interchangeably between consultant and construction firms for generations. The control is purchased at the price of payment for instructions and the meeting of hazards on a relatively generous scale. Many of the arrangements flow over into the present contract and it may be expected to be used in a similar way, tempered by the lump sum concept which appears to be at its heart.

There is no indication of the scale of works intended for this contract and no particular limit needs be postulated. The larger the works, the less likely are they to be susceptible to predominantly firm design and the greater the risk in attempting it.

Contract basis and documents

The contract is of a lump sum nature, so differing from the ICE standard contract and aligning with the JCT contract. This aspect and some differences within it are taken under 'Contract price' below.

Relationship of contract documents

The contract is defined by clauses 1(1)(e) to 1(1)(h) as having several elements, which may be rearranged for comment:

(a) The conditions of contract.
(b) The employer's requirements, which may be expected to be equivalent to those under the JCT contract, but which are specifically defined as those at the award of contract and as 'including any subsequent variations thereto'. The latter expression could be difficult to interpret if pressed in the context of some clauses of the contract.
(c) The contractor's submission, which again may be expected to be equivalent to the contractor's proposals under the JCT contract and which again includes documents as agreed prior to the award of contract (a fine distinction may be noted here).
(d) The submission also includes 'the tender' and there is a form of tender included within the contract binding, to which is annexed an appendix containing items similar to those in the JCT appendix, such as contract dates.
(e) The contract price is 'the sum to be ascertained . . . in accordance with the Contract'. The form of tender states that the contractor will 'design, construct and complete the said Works' for 'the sum [inserted] or such other sum as may be ascertained'. This appears to give a lump sum price subject to adjustment, and not a tendered amount set aside for the usual civil engineering complete remeasurement.
(f) The written acceptance of the contractor's submission.
(g) The contract agreement, there being a form of agreement equivalent to the

JCT articles included within the contract binding. This does not record the contract price, but relies on the contractor's submission.

(h) There are also 'such other documents as may be expressly agreed' and these should be given in the form of agreement in or by reference from the space under 'the following documents'. As commented several times in discussing the JCT contract, most documents should be embraced within what here are the requirements or submission. Within the contract binding there is a form of bond which is optional: this needs mention here.

There are also definitions of 'prime cost item' and 'contingency', the latter being in terms making it a provisional sum in JCT terminology. The former is something not envisaged under the JCT contract and is mentioned further under 'Contract price' below. It is left open as to where in the documents either may occur.

The status of contract documents is given by clause 5(1)(a) as 'mutually explanatory of one another'. This means that no immediate precedence order is established, although the broad priority of the conditions may be assumed, as is done by accepting clause 5(1)(a) itself.

Ambiguities and discrepancies

These elements relate to more than, but particularly affect, design. The terms used in this contract may be taken for present purposes as covering essentially the same ground as discrepancies and divergences in the JCT contract, that is a range of differences.

Clause 5(1)(b) provides that if there are differences between the requirements and the submission, the former are to prevail. As with the GC/Works/1 contract, this reverses the position which is not explicit but may be deduced under the JCT contract, where the 'appears to meet' provision of the recitals applies. In practical terms, here again the contractor must ensure that, while his submission may interpret and develop, it does not amend or conflict with the requirements. If he wishes to achieve this effect, he may seek direct amendment of the requirements (as is contemplated by these conditions) before the award of the contract. Alternatively, this contract expressly permits him to ensure that the position is clear 'in the light of the several documents', as clause 5(1)(b) has it, that is that there are specific statements over how the differences are to be interpreted. This latter path may be the more expedient one to pursue when involved interlocking or overlapping provisions exist, but this is also when it becomes more difficult to explain what is meant rather than to remove the conflict.

This feature means that clearing differences entirely within either one set of documents is also affected. In the case of the requirements and as they prevail, differences in them have to be 'explained and adjusted' under clause 5(1)(c) by the employer's representative. The contractor cannot be held effectively to have made his own choice of what applies, as occurs under the JCT contract where the proposals prevail and where the employer has to issue countervailing instructions to secure something else which he may actually want. Under the GC/Works/1 contract, it is stated that resolution will produce a change, so assuming that the contractor has *not* chosen the now desired option. Here the representative has to issue clarifying instructions, with the contractor receiving extension of time and reimbursement of cost, but only when 'beyond that reasonably to have been foreseen by an experienced contractor', inclusive of a profit allowance. If the requirements are ambiguous, it is difficult to see in principle how even the most experienced contractor might always know when tendering what cost he should reasonably foresee. The provision is unclear, if not itself ambiguous, and an experienced contractor ought to raise any noticed problem when tendering.

In the alternative situation of differences entirely within the submission, clause 5(1)(d) simply states that these are to be 'resolved at the Contractor's expense'. This leaves open whether the employer may choose which version to accept or whether the contractor may put forward another solution, as the JCT and GC/Works/1 contracts allow. On the *contra proferentem* principle, it would be reasonable for the employer to have the benefit of any doubt.

Final operation and maintenance

Before issue of the certificate of substantial completion, clause 48(3) requires the contractor to provide 'operation and maintenance instructions'. Clause 61(3) follows this with the provision of draft and final 'operation and maintenance manuals' before the issue of the defects correction certificate. These allow the employer's representative the opportunity to be sure that the information is adequate and this is a very useful arrangement. It is best operated when there are clear statements in the employer's requirements initially.

These documents are not 'as built' drawings and stipulations about any such need to be given in the requirements.

Contractor's design responsibilities

The provisions falling within this category are generally fairly brief until matters relating to ground conditions intrude. This is to be expected for works of civil engineering, but issues may be noted which may be applicable to building substructures and warrant mention in the employer's requirements.

Insurance

There are no provisions about design insurance, after the manner of the JCT contract. Insurance of the works is limited to that against damage and so forth during progress. Among the excepted risks against which the contractor does not have to insure under clause 20(2) are 'any fault, defect, error or omission in the design of the Works for which the Contractor is not responsible'. This is the only mention of design and insurance together and is interesting in passing in that it recognises the possibility of design responsibility lying elsewhere, perhaps with the employer.

Nevertheless, the contractor should clearly hold adequate design insurance here just as much as for any other type of works.

Design of the works

The contractor's main design responsibility is given in clause 8(2), following his general obligation to 'design, construct and complete the Works' in clause 8(1) and providing everything necessary in the process. It is a responsibility to 'exercise all reasonable skill, care and diligence', so equating with the JCT responsibility but without spelling out any such parallel standard as that of a professional designer.

The responsibility goes further though, by requiring the contractor to check any design emanating from the employer and included in the requirements and then to accept responsibility for it, subject to any modifications agreed between the parties. This does make clear what is left open by the JCT contract, but such checking may be a highly onerous matter to do during the usually short time of tendering. It is limited to 'design ... included in the Employer's Requirements' and this would appear not to cover work designed by or for the employer, but for which the design information is not included sufficiently for checking. A drawing of a retaining wall, for instance, would be sufficient if it gave all concrete and other dimensions and sizes and dispositions of reinforcement, related adequately to ground conditions and other factors. But the contractor would not have to check and be responsible for the soil survey data and this would limit his responsibility for the dependent design. The next heading takes this issue further.

Matters related to design

The main question of design by the contractor is preceded and followed by several related issues. There is provision in the appendix to the form of tender for the contractor's designer to be named and, by clause 4(2)(a), the contractor has to obtain the employer's consent to any change in designer. This is an important matter and such a provision could well be in other contracts. No provisions relating to novation of the employer's designers are given.

Clause 6(1) obliges the contractor to give notice of 'any further information' he requires for the 'design and/or construction of the Works', with a rider that he may have remedies over time and payment if the information is delayed. This is a little odd in a contract in which the contractor has already tendered a definite price for the works and must surely refer only to such subsidiary instances as when there are related contingencies or provisional sums in the contract originally.

By clause 6(2), the contractor is to let the employer's representative have 'designs and drawings ... to show the general arrangement' and compliance with the requirements and construction work may not commence until the representative has given consent. If the representative considers that there is a clash with the requirements, he may withhold consent until the contractor makes 'appropriate modifications'. There follows a provision about the contractor wishing to modify his design further for his own reasons. Again, if there is delay by the representative, the contractor has remedies over time and payment. Subsidiary provisions relate to supply of documents and copyright in information.

Clause 14 deals initially with matters of the programme, but ends unexpectedly with matters of design. In clause 14(5), it requires the contractor to provide the employer's representative with 'design criteria relevant to the Employer's Requirements' (but not further). By clause 14(6), this information is to demonstrate that the permanent works will not be adversely affected by stresses etc imposed during construction and is to include matters relating to the contractor's temporary works and equipment, the incidence of which is generally much higher in civil engineering.

Importantly for present and other considerations, clause 1(7) provides that consent or approval given to the contractor is not to relieve him of any of his obligations, particularly over 'his duty to ensure ... correctness, accuracy or suitability'. A similar provision is in the GC/Works/1 contract and the JCT contract's supplementary conditions and could well be included in the JCT contract's main conditions to guard against ill-conceived actions by the employer or his agent.

Site conditions

The way in which this contract deals with the site is significant and is due to the heavy incidence of groundworks in civil engineering, while it impinges particularly on design. There are provisions in the GC/Works/1 contract which allow some of the main elements here to be handled reasonably smoothly. The underlying principles of both could well be taken into account in bringing additional stipulations into the employer's requirements under the JCT form when earthworks are a major and risky consideration. This will often be fairer and avoid the possibility of tenders overloaded as a precaution when the contractor is asked to carry all ground risks.

By clause 11(1), the employer is deemed to have made available to the contractor for tendering purposes 'all information ... obtained ... from investigations undertaken relevant to the Works' over two categories, of which the more important here is 'the nature of the ground and subsoil including hydrological conditions'. This imposes no responsibility on the employer to obtain any such information, but there is a duty of disclosure and an assumption of responsibility for the accuracy of what is given. However, the contractor is expressly made responsible for interpretation of what is given and so for how he chooses to use it in the design and execution of the works and the resulting costs.

The liability of the contractor is taken further in clause 11(2) to cover proper inspection of the site and of 'information available in connection' which stretches beyond what the employer provides under clause 11(1). This liability is left to be implied in the JCT contract, although commonly it is included in the preliminaries of any bills or specification. The difficult element is 'the form and nature thereof including the ground and subsoil' which could affect the question of design radically, as may be seen from the example of a retaining wall given above. The liability is qualified by 'so far as is practicable and reasonable before the award of the Contract'. This might be held to mean that no ground exploration work could be performed by the contractor during tendering, but he is always advised to make this point with his tender and also that he is not able to do any such work in the further interval before actual award of the contract. Clause 18 taken below allows for its performance after work has started on site and bears further on the question of responsibility for cost.

As a generalised statement, clause 11(3) provides that the contractor is 'deemed to have based his tender on the information made available by the Employer and on his own inspection and examination all as aforementioned'. This includes the common implied or express point that the contractor cannot plead ignorance of what he should prudently have found out and so the clause records that he should have taken account of what has been provided. But it is a two-edged sword in that the employer is effectively responsible for any information which he has supplied, subject to the questions already discussed of whether the contractor has interpreted information properly and checked design when possible.

Clause 12 deals with 'Adverse physical conditions and artificial obstructions' and envisages possible additional payment and extension of time when these 'could not ... reasonably have been foreseen by an experienced contractor'. 'Adverse' usually means unduly hard, soft or wet subsoil conditions. Under other civil engineering contracts, the payment matters are often covered in the measured items allowed, both for excavations and any redesign of the permanent works, probably resulting in more extensive or stronger structures. In the present contract the contractor has to deal with design and it may be argued that the resulting extra work falls into the category of 'not ... reasonably ... foreseen' and so ranks for payment. This perhaps results from the straight transfer of the clause from its original setting, but is unclear. It is as well for the matter to be dealt with in the employer's requirements to make absolutely plain which party is to carry the risk on the particular project.

Clause 18 allows the contractor to undertake boreholes or other ground explorations during progress with the permission of the employer's representative, who will have regard to the permanent effect on the ground conditions. Copies of all data obtained are to be provided to the representative, who presumably may make them available to the employer for future reference. In general all this work is to be at the expense of the contractor as part of his design development cost and perhaps also having a relation to his working methods or temporary works. The position is qualified by a statement that the work is to be paid for as extra when it is due to conditions or obstructions under clause 12 or a variation under clause 51.

Financial provisions

This contract is somewhat vague by comparison with the two forms bearing direct comparison about setting out its financial basis. The JCT contract has a contract sum clearly stated to be adjustable in accordance with the conditions, but otherwise fixed, and even spells out somewhat repetitiously the elements of adjustment. The ICE standard contract in its form of tender refers to 'such sum as may be ascertained', so that there is no total sum acting as a contractual point of departure, although the total, and probably the detailed pricing, of the original approximate bills of quantities forming the basis of tenders will have been a major factor in deciding which tender to accept. It then has clauses leading to complete remeasurement of the approximate quantities, while giving authority for variations of design and stating that simple fluctuations in quantities without design changes do not need instructions to validate them.

In this contract, there is some adaptation of that wording. The form of tender contains an offer to perform 'for the sum of . . . [to be inserted] . . . or such other sum as may be ascertained', but does not refer to the original sum by any title. The 'Contract Price' is defined in clause 1(1)(h) as 'the sum to be ascertained and paid', which *appears* to equate with the 'such other sum' of the tender. Again there is authority for variations in design in clause 51 and for their valuation in clause 52, similar to those in the other contract, but including a simple provision for quotations from the contractor as a first basis for agreement of the valuation, given optionally in advance of the work.

There is no provision amounting to authority for remeasurement of the whole works and further there are no comprehensive bills of quantities upon which to base this. The corresponding clause 55 has a quite limited provision that 'Where the Contract includes any bill of quantities schedule of works or Contingencies the quantities set out therein shall be deemed to be the estimated quantities but they are not to be taken as the actual or correct quantities of work to be constructed'. Clause 56 has a provision about attending to measure introduced by 'If . . . it becomes necessary to measure any part'. These two clauses are not expressly linked, but most reasonably imply that restricted parts of the works are to be measured as executed, although another interpretation could be placed upon them.

The financial scheme of the contract therefore appears to be:

(a) An initial total amount, unnamed but corresponding to the JCT contract sum as being fixed and covering the works as designed by the contractor and put forward in his submission in response to the requirements.
(b) Within this amount, sums which are provisional, relating to subsidiary elements of the works and based upon approximate quantities within bills, schedules or contingencies (effectively equivalent to provisional sums under other contracts). This does not exclude the possibility of any of these bases including or consisting entirely of lump sums, also provisional in amount. All of the original sums are replaced by the remeasured or otherwise valued final sums.
(c) Also within this amount, prime cost items which do not relate to any formal system of nomination but would suit such an approach, there being allowance for the addition of overheads and profit at the percentage given in the appen-

dix. (Provisions for domestic sub-contracting of a regular nature are given in the conditions and may be of use here for minor packets of work. Strictly, they cover construction only and not design.)

(d) Provision for variations and other adjustments to be made.

This scheme is quite normal, but its essence has to be derived from assumptions of normality and common practice. It is desirable for the position to be made clearer in the employer's requirements by definition of the elements actually being used, or even for the pattern to be varied in some of its detail to suit particular needs. It needs some comment by comparison with the JCT approach.

There is no mention of a contract sum analysis, but it will usually be an indispensable document, both for interim payments and valuation of variations. Such a document has been discussed in some detail under the JCT contract and an equivalent may be deduced to suit the differing work here. For the reasons given there, the analysis should consist principally of lump sums rather than quantities. However, those parts of the analysis given as (b) above are often likely to consist of approximate quantities to support remeasurement for the particular reasons given below. These elements may then have a wider use in pricing some parts of other variations. The wording in the GC/Works/1 contract suggests an intention to use some quantities there on occasions.

The use of sums which are provisional parallels the JCT contract, but there is no statement here as to whether their inclusion is restricted to the employer's requirements, as happens there. The use of prime cost items is peculiar to the present contract and again their inclusion is not restricted to one documentary location.

This openness is more understandable here, as civil engineering works often lead to greater initial uncertainty over what will be required. This may be a matter of the straight quantity of a fairly inevitable type of work, such as mass concrete, but may be one of more radical doubt about the nature of work, or even whether items like sheet piling or dewatering will be necessary at all. The contractor naturally has every reason to limit his risk and may wish to make a provisional allowance to this end. This is a different situation from that in building work (at least in superstructures) when the contractor may choose relative risk by not designing fully to meet a defined contingency in advance of obtaining the contract and so may make a covering financial allowance against what is in principle knowable at the time.

The employer should balance the uncertainty of a provisional allowance put forward by the contractor in his tender against the certain, but possibly higher, allowance which might otherwise be made to cover the risk situation. This may render it desirable to allow inclusion of the contractor's sum in his tender. If so, there is the question of whether the inclusion gives the particular tenderer an edge, that is whether parity of tendering has been eroded and other tenders should be reviewed and adjusted. In any case in which a tenderer has put forward his own approximate quantities and firm prices, it is necessary to assess whether they are both a reasonable allowance against the anticipated work.

In the case of all provisional allowances, but especially those subject to remeasurement of quantity against fixed unit prices, the problem of quantity control occurs. This is mentioned in looking at design changes in Chapter 9. It is sometimes more acute in civil engineering because of the scale of single elements of the works. It is that the contractor does not have the same incentive to economise in design when a variation is involved as when he is designing as part of his tender. If he designs a retaining wall to be unduly thick to reduce his design risk, he is prima facie due to be paid for the greater quantity of concrete placed, measured and priced in accordance with what is in the provisional contract allowance. The employer's representative may query the design, but this has to be done carefully to avoid implication if it turns out to be inadequate, so transferring responsibility from one party to the other.

Mention has been made of clause 6(1) where the contractor is to obtain from the employer 'any further information ... required for the design and/or construction'. It is presumably not intended that this should herald what are effectively variations, but is the sort of provision which well might lead to them and the contractor should be alert to this possibility.

Chapter 17

Performance specified work

Background considerations

Pre-contract arrangements and design

Post-contract design and execution

As noted in Chapters 11 and 12, there are two common ways of obtaining design and specification of some part of the works within a main contract which is mainly a build-only contract, without the necessity of isolating the work in a nominated sub-contract, which also may depend on performance specification and has numbers of principles corresponding to those in this chapter (see Ch. 18). One is by using an arrangement such as the JCT contractor's designed portion supplement considered in Chapter 12 when the extent and separateness of the portion are quite high. The other method is that of introducing performance specified work, as mentioned in Chapter 4, which is taken in this chapter and is intended for relatively limited work. The methods are not mutually exclusive and they may be used together in the same contract. The situations for using one or the other are not conceptually exclusive either, as may be inferred from the terms 'quite high' and 'relatively limited' used in this paragraph.

Background considerations

In the past it has been common practice to use some such expression as 'contractor-designed construction' to identify the work now being considered. This was usually introduced by the method of measurement supporting the bills of quantities. The work was measured in accordance with the rules given and whatever was necessary to specify its performance characteristics was given in the description of the item or items. For example, in some earlier editions of the Standard Method of Measurement of Building Works reinforced concrete slab construction was so given and, typically, spans, loadings and support spacings would be stated and, possibly, limiting thicknesses and soffit appearance. The quantities were accurate, but the descriptions left it open to the contractor to develop technically what was required. There might be slight inaccuracies introduced into the quantities for adjacent items if, for instance, the actual thicknesses of floors changed either the heights of partitions or the height of the envelope of the building. The approach is also applicable to roof decking, windows, curtain walling, mechanical plant and much else. A quantitative approach in a thoroughgoing design and build contract could possibly follow such a route for a contract sum analysis, although this is likely to lead to problems, as is considered in Chapter 9.

It is common to find bill preambles for such work which in effect say: 'The contractor shall be responsible for performing all detail design within the performance criteria specified, for obtaining all statutory approvals and for supplying copies of drawings, calculations, specifications etc to the architect for approval (which shall not remove the contractor's responsibility) before any work is put in hand.' Such stipulations tend to cover responsibility in the sense in which the term is used in Chapter 5, that is responsibility for performing the various contractual duties. Unfortunately, they have often failed to deal at all with the design liabilities arising when performance is inadequate and, if they have attempted to do so, may well have been suspect in their relationship with the contract conditions, which also have been extremely reticent on the issue. Further, the building method of measurement which blithely gave the presentation rules was also fortunately dumb on the question of liabilities.

Until recently, the contractor's designed portion supplement has been the only standard document widely available, if not widely used, in the role of sledgehammer for cracking the nut of small parcels of work and somewhat unsuitable for work readily quantifiable and intimately related to other contiguous construction. The JCT standard form of building contract in all of its six editions now has clause 42, entitled 'Performance specified work', dealing with the matter. In doing so, it goes into the broad procedures, but not into what may literally be the nuts and bolts of the design and its presentation, so that the type of preamble given above may still be relevant, so long as it is not in conflict.

A sense of proportion is needed in endorsing work as performance specified work, so that for small or non-critical items it is best if nothing is done. It is after all usually the electrician who designs final sub-circuit layouts and the joiner who chooses his own nails! In the case of components, suppliers are likely to be under a 'fitness for purpose' obligation, which the courts are tending to import more into site works (see Ch. 5). Ironically, this could give a greater depth of liability than using the various 'skill and care' contract provisions, provided liability could be shown to run at all.

In what follows, matters are described specifically by reference to JCT clause 42 in its with-quantities edition (the operating effect is the same under the other editions) and so to inclusions in bills of quantities. Occasional references to the contractor's designed portion approach in Chapter 12 are usually given simply as to 'the designed portion'.

Pre-contract arrangements and design

The aspects of briefing, competition and design differ in several ways from the corresponding elements for the designed portion. These reflect the lesser status of performance specified work within the whole and the reduced opportunity for design development it usually offers to the contractor. The architect is likely to have made such decisions as the configuration of the structure as a whole and the major materials to be employed for the part now being considered.

For instance, a roof may have been designed to the stage where it is determined to be sloping to a designated pitch and profile rather than flat, and where the coverings are to be a given sheet metal on a given sub-structure: the contractor may then be told that there is to be a steel roof framework with purlins and trusses and that the steelwork constitutes performance specified work, but that what it supports or what supports it does not. As mentioned above, the designed portion approaches and performance specified work are not mutually exclusive and this is illustrated by this example. Even if the roof coverings also were of the contractor's design, it would still be possible to use the performance specified approach.

Depending upon the nature of the work, it might be possible to extend the contractor's discretion over his design to the specification of materials to be employed. Thus it could well be straightforward to do this with the metal roof coverings, if some latitude was permissible on functional grounds. It is less likely that the same discretion could be offered over the supporting framework; if it could, it might allow choice between various metals, but might not extend to laminated timber instead. It might also be difficult to allow variations in the spacing of such items as purlins.

If the whole roof were a clear-cut entity above a common datum level and the contractor had flexibility over the choice of materials, and even more over the roof shape, the balance might well tip in favour of the designed portion approach.

So far as briefing is concerned, the contractor is faced with a final statement of performance specified work in the tendering documents, with fairly limited choice over what he may propose. There may have been earlier discussion over the type of work included, and advice on design, and perhaps price, is likely to have been taken from any specialist trades concerned. Anything akin to two-stage tendering by the several contractors will not occur, as it just might with difficulty for a designed portion.

Clause 42.1 requires performance specified work to be identified in the contract appendix, to be provided by the contractor and to have 'certain requirements . . . predetermined . . . and shown on the Contract Drawings'. 'Certain' carries its usual overtone of uncertainly variable in principle, but definite for the particular project. The requirements given do not preclude sub-letting and clause 42.17.2 provides for this by stating that the contractor's responsibility is not diminished if it occurs. There is no bar on using the general arrangements of clause 19.3, which cover a list of at least three persons (effectively meaning firms) given in the bills and from whom the contractor is to select a sub-contractor. This is a very useful way of channelling the contractor's choice and keeping contact with any specialists with whom the architect has been dealing to obtain preliminary advice. It has the disadvantage that the main contract tenderers are all likely to be known to the persons on the list of three, but this is no novelty.

However sub-letting occurs, the optional domestic Sub-Contract DOM/1 has clauses stepping down the provisions of the main contract to suit. On the other hand, the obligatory nominated sub-contract conditions NSC/C do not provide for performance specified work at either main contract or sub-contract level, as the total pattern provides for design in other ways (see Ch. 18).

Clause 42.1 also requires performance specified work to be included within the bills by one of two means. The first, and hopefully that regularly used, is by the inclusion of quantified items as already mentioned. There are no provisions in the seventh and current edition of the Standard Method of Measurement of Building Works (SMM7) for performance specified work by that or any other name, but the rules of measurement and description given there will often suit the work concerned, perhaps with some qualification over points where the contractor's design decisions may vary details. If necessary, special rules should be devised to suit the work and deal with the treatment of any quantity adjustments introduced by detailed design following quantification. Clause 42.9 accommodates this approach in principle by stating that the inclusion of performance specified work does not constitute a departure from the method of measurement stated in clause 2.2.2.1, which applies to the works in general.

The alternative method of including performance specified work in bills is by a provisional sum. Here, clause 42.7 lists information which must be given in the bills, even though the sum is adjustable to relate to the specific work required. One item is the performance required of the work, which repeats the essence of what is in clause 42.1.4 and establishes a broad level of responsibility to be assumed by the contractor. The others are the location of the work and adequate details for assessing programming and pricing preliminaries items, so that the contractor is aware of how his work at large is affected and so that the contract sum is not deficient. The provisional sum is close to one for defined work (as given in SMM7) included in bills of quantities apart from any performance specified work. This arrangement gives a fair guide over what is to happen, subject to a lack of immediate technical fullness and to price uncertainty, which is a standard disadvantage for the employer when using provisional sums. The disadvantage is often stronger here, as the work is usually by definition not susceptible to pro rata pricing against other work in the contract and includes the extra facet of design where the contractor holds the initiative.

Unlike when the designed portion is in use, the contractor is not required here to provide anything equivalent to contractor's proposals at the time of tendering. This reflects the relatively subordinate nature of the work, with the contractor completing technical design and specification within a closely defined scheme and not able, for instance, to change matters like room planning and fenestration. He simply prices the work as presented in the bills as part of his tender and based upon his intended technical development, as he does for the whole of the works under the BPF system (see Ch. 13). In an important case, such as when the choice of contractor is at stake, the architect should seek information with the tender, but formally everything is done post-contractually.

Post-contract design and execution

There are several contractually defined elements in carrying out performance specified work within the works:

(a) The contractor is to give a statement under clause 42.2 providing documentary details before performing the work.
(b) The statement is to give adequate information by clause 42.3 on 'the Contractor's proposals' [*sic*], including whatever was specified in the contract bills or in the instruction to expend the provisional sum. Such information may or may not extend to drawings, perhaps influenced by their usefulness as 'as-built' information.
(c) The statement is to be provided under clause 42.4 by the date given in the bills or in the instruction over the sum, as the case may be. In the absence of a stipulated date, the contractor is to provide the statement a reasonable time ahead of the work. Prudently, the bills could give the time as a margin rather than as a calendar date, to allow for the contractor's freedom to establish and modify his own programme.
(d) The architect may require the contractor under clause 42.5 to remedy any deficiency in the form or detail of the statement, that is to render it clear and complete. The architect's action or inaction here does not relieve the contractor of responsibility for his proposals.
(e) The architect is to draw the contractor's attention under clause 42.6 to any deficiency in the effective content of the statement 'which would adversely affect the performance'. By inference, the contractor need not take any specific action and the architect certainly should not name any. Again the contractor is to remain responsible and the architect should not go beyond ensuring that the statement measures up to the 'appears to meet' criterion.
(f) The architect is to give any instructions under clause 42.14 necessary to integrate the performance specified work with the rest. If his original design took precise account of the configuration etc of the performance specified work, this should be unnecessary. It may arise if the contractor has come forward with some proposed modification, such as in the spacing of members making up his work and needing support. This allows either part of the work or both to be varied and will lead to a variation to be valued as usual, which is a distinction from the position under the designed portion arrangement where no financial adjustment takes place. There is room for debate here if the contractor proposes something leading to a variation addition in the rest of the works, when it might appear that he could have obviated the extra by proposing a dearer version of his design.

(g) By way of counter, the contractor may notify the architect under clause 42.15 if he considers that compliance with the architect's instruction will 'injuriously affect' the performance specified work, this term being that also used in the designed portion and discussed in some detail. The matter is left effectively for the two to come to a reasonable agreement.
(h) Should the contractor delay in providing or amending his statement then, by clause 42.16, he is not entitled to remedies for delay by way of extension of time, loss and expense or determination. Under the provisions of the conditions generally, he will have such remedies if the architect is tardy in his actions.

These provisions regularise what would usually emerge as a working procedure in most cases, while highlighting several places where strains may occur. The intervening clauses 42.8 to 42.12 cover several provisos clarifying matters not needing comment here. The whole set may be compared with the more detailed stipulations in the designed portion, especially those summarised in Table 12.1 and over discrepancies. The work proceeds along with the rest and the architect exercises his usual powers of supervision and inspection, covering compliance with the design and specification and all matters of quality. Two points remain.

Clause 42.13 requires the contractor to 'provide an analysis of the portion of the Contract Sum' for the performance specified work. This is to be done 'within 14 days of being required . . . by the Architect' unless, as is better, an analysis has already been given in the contract bills so reducing any question of late 'variation spotting' by anyone seeking to allocate values to their advantage. If the works arise out of a provisional sum, the analysis can only occur after the instruction and may be the least of the financial conjurings. The form of the analysis will be determined if it is given in the bills. If it is not, it is still desirable for it to be given to the contractor in advance of his preparation, so that it is suitable for all contract and related purposes, some of which he may not know.

Clause 42.17.1 places a simple 'reasonable skill and care' responsibility upon the contractor over the 'provision of Performance Specified Work', a term apparently embracing both 'design and build' aspects. However, there is a rider that the general obligations about materials and work are not affected: these give a strict care liability in these areas. The clause does not add any of the provisions which occur in clause 2.5.1 of the JCT with contractor's design contract about liability equating with that of 'an appropriate professional designer', presumably as there is not the same question of scheme-based design. This is closer to the liability statements given for nominated sub-contractors (see Ch. 18) where, as in the present case, the architect's integrating role is more dominant.

Chapter 18

Nominated sub-contracts

Arguments for and against nomination

Design and nomination

Scope of and procedures for design etc

Subsidiary provisions over design etc

Like some of its predecessors, this chapter is concerned with selected aspects of the system that it considers. Nominated sub-contracts have been introduced and the standard documents listed in Chapter 4. They may be used within some JCT main contracts for which the architect provides the design under direct appointment by the employer and in which the contractor and all his sub-contractors, domestic and nominated, construct the works in accordance with that design. It is, however, possible to arrange for a *nominated* sub-contractor to design his own work in whole or part and to be responsible to the employer for this design, rather as when performance specified work is included in the contract for the contractor to price directly and for which he assumes responsibility over the design (see Ch. 17). However, as is noted hereafter, there is not the same detail in the provisions over this as occurs for performance specified work and the matter is approached contractually rather differently.

This chapter is concerned principally with those aspects of nominated sub-contracting which concern design, but the general advantages and disadvantages of nomination are set out first to give a balanced view over this one issue.

Arguments for and against nomination

Principles

Under nomination, the architect puts forward a person of his choice whom the contractor must accept after the main contract is awarded as a sub-contractor, subject to a right of objection to the person. This enables the architect to select someone of the right experience and quality to perform specialised work within the contract. Because the contractor has only a limited, negative control over selection, the price payable to him for the works is adjustable according to the amount of the sub-contractor's account. This is facilitated by instructing the contractor in the bills or specification to include in the contract sum a fixed amount, known as a 'prime cost' sum (because it represents the amount payable to the sub-contractor himself), for the sub-contract work. No detailed quantities or specification are then provided in the main contract tendering documents. The architect obtains the sub-contractor's tender, usually with the aid of the quantity surveyor, and instructs the contractor to place the sub-contract in that sum. When the work is performed and the sub-contract final account agreed, based on the tender, the amount of this account is added into the contractor's final account in place of the prime cost sum. The contractor's profit and cash discount allowances, but not his attendance allowances, vary in proportion to these adjustments.

The financial basis for the sub-contract may be a simple lump sum, or one based upon firm or approximate quantities or upon a schedule of rates for measured work. It is unusual for it to be based entirely upon cost-plus reimbursement, also known as 'prime cost', although incidental daywork may be included. It is unfortunate that the industry uses the term 'prime cost' to indicate the two quite distinct financial mechanisms mentioned in this paragraph and that immediately preceding.

Criticisms

While nomination secures an immediacy of quality control for the architect, it is often criticised justly or unjustly, and a balance must be struck over whether to use it in a particular case. A common criticism is that the architect does not obtain as low a price as would the contractor for the same work. This may reflect the contractor going to a different range of firms, and then brings in the question of comparability over quality. But it is certainly true that on occasions, although not invariably, the contractor can obtain lower terms from the *same* firms. This may be a simple fact of difference in commercial muscle, depending on the scale of work offered annually by the particular architect or contractor. It sometimes indicates that the contractor is using selection techniques which depend less upon considerations of absolute parity, and more upon reiterative procedures, better known as 'Dutch auctions'!

The other major, and louder, criticism is that the contractor does not have the same control over a person nominated to him, as over one of his own choice. There is no doubt that, contractually, the nominee *is* a sub-contractor responsible to the contractor, who in turn is responsible for him to the employer. Both the sub-contractor and the main contract establish this, so that the architect can deal with the sub-contractor only through the contractor over matters like quality, performance and instructions, while even settlement of his final account by the quantity surveyor (if involved) also occurs strictly through the contractor. In the nature of the sub-contractor's specialised work, many of the informal discussions on these aspects that precede the formal dealings are usually conducted directly with the sub-contractor, often to the contractor's great relief! The temptation for the architect is then to take short cuts in procedures and not use the proper written channels of communication. The corresponding temptation for the contractor is to allow this to happen, and then to blame hiatuses in running the job on such short-cut methods.

In addition, while the nominee remains a sub-contractor, he does have distinct treatment under the main contract over several aspects, especially extension of time and interim and final payments, which is not afforded to other sub-contractors. Again the presence of these provisions does not disturb the sub-contract relationship, but they do create an air of distance, so that the contractor may feel (improperly) a slackening of responsibility. Some of these provisions are referred to in and affect the employer/nominated sub-contractor agreements discussed in this chapter.

Alternatives

An alternative to nominating sub-contractors is to employ specialists under direct contracts with the employer, alongside but with no contractual relationship to the main contractor. This is a possibility recognised in, for example, the JCT with-design form and standard forms in clause 29. It means that the employer, or his architect, controls the persons directly, while the contractor merely allows them to work on 'his' site. The employer or architect therefore has to deal with overall co-ordination.

This may be a useful approach at two extremes: a small, self-contained element performed late in the programme and not needing close integration with the bulk of the works, such as a mural, or an installation disproportionately large physically or financially in relation to the building contract. It also means that the employer does not pay cash discount or profit *as such* to the contractor, who will include elsewhere in his contract sum for what he requires as the equivalent and to cover accommodating the specialist's activities. Against this, the employer may face more extra payments when co-ordination is inadequate. The system suits the design situation well, as there is contractual privity. The more integration there is, the riskier does it become for the employer, who becomes his own 'management contractor' when it is used extensively. It is therefore employed fairly occasionally – on most projects not at all.

The main alternative to nominating sub-contractors is to give full quantities or specifications to the contractor when tendering, and to ask him to price the specialist work in detail and in competition. It may be stipulated that he is to sub-let to one of a list of permitted domestic sub-contractors, or it may be his discretion whether to sub-let at all. If he wishes to do so in the latter case, he must obtain the architect's consent to the person proposed. Both of these occur in the JCT standard forms, while the JCT intermediate form has a procedure for naming a *single* sub-contractor with the work priced by the contractor. However sub-letting occurs, the contractor is paid his own tendered price (subject to normal contract adjustments) for the work, whether or not he based his price on any particular sub-quotation, and whoever he actually uses for the work. This method still affords some control over the person who is to perform the work.

In the case of an intermediate form named sub-contractor, the architect has obtained sub-contract tenders ahead of the main tenders and has selected the tender of the named sub-contractor so that all details, including price, can be made available to the main tenderers. An element or the whole of the design may have been performed by the sub-contractor and in this respect the system operates rather like nominated sub-contracting being described here, with a collateral agreement to link employer and sub-contractor.

The problem with the domestic approach may be precisely that it does mean giving full quantities or specification at the possibly early time of the main contract tendering period. This means extra work at that stage and perhaps earlier decisions, although many of these are what the JCT contracts assume. Procrastination may be a failing, or it may enable later technical developments to be taken on board. It may be noted, by way of comparison, that the contractor is expected to make these early decisions in a design and build contract, or take the risk that he can absorb his later decisions into his design and price.

Underlying this problem may be the reason for using nomination, rather than the domestic approach, which leads to it being mentioned in this book. This is the desire to obtain design of the whole work or some part, or a development of detailing from the specialist concerned. The domestic sequence is: design, specification and any quantities, tendering, selection of sub-contractor. Design therefore cannot be undertaken by the sub-contractor, while early advice by him may prejudice later competition by locking decisions into his techniques or components. Further, the contractor has no liability over design matters and so cannot accept any over design by his sub-contractor.

One answer to the problem, provided by the JCT standard form contract, is to use the performance specified work arrangement as given in Chapter 17, when sub-letting is available. Another is nomination as being discussed, which opens the way to design by the sub-contractor and to a commitment over liability. It also, as distinct from performance specified work, allows some decisions (perhaps many) over design of the sub-contract work to be deferred until some time into the post-contract situation. With prices being obtained nearer to the time of the work, it is also possible to obtain more realistic tenders, to the benefit of one party or the other.

Design and nomination

Introducing design

The possibility of design by the nominated sub-contractor arises because he is selected by the architect, rather than the contractor, so that the architect may take in design ability as one criterion. If so, he may brief one person or more to produce a 'scheme and a price' for a defined part of the works and based upon aesthetic, performance or other stipulations. This procedure leads to what is effectively a design and build (or install) sub-contract, although it is not usually called this.

Much of what is said about design and build main contracts in this volume is just as applicable here. The main difference is that the sub-contractor's work is not self-contained, but has to integrate in appearance, performance or in other ways with other parts of the works. This means that the architect, and perhaps before him the employer, has to be very careful in defining boundaries and ensuring that standards set are complete and unambiguous. Otherwise, the sub-contractor may be in doubt or may be misled and, either way, if he does not query the position or if he makes incorrect assumptions, he may produce a design that may not be found inadequate or otherwise unsuitable until too late. While the employer/nominated sub-contractor agreement mentioned below contains stipulations about performance specifications, it does not give the architect power over integrating work post-contractually as under the contractor-designed portion supplement discussed in Chapter 12. As a result, integration must be introduced either before acceptance or by way of a variation, so leading to a difference in payment, probably upwards. The payment aspect is not mentioned in the standard architect's conditions of appointment which state the he 'will have the authority to co-ordinate and integrate such work'. This is an expression to the employer only, amounting to 'all will be well', but in itself not binding on the sub-contractor.

The JCT tender, nomination and sub-contract forms are quite silent about design and its integration. They are for work of constructing or installing what has been designed already, with the design supplied to the sub-contractor by the architect. That he may obtain the design from the selfsame sub-contractor, approve it (with similar limits to those considered in Ch. 5 over design and build proper) and then issue it to the contractor with an instruction for the sub-contractor to proceed, is no concern of the sub-contract. This accords with the lack of design responsibility in the related JCT main contract. Design is something happening 'out there' on the far side of the architect from the contractor. Variations may be instructed under the sub-contract, through the contractor, but they should originate with the architect and result in the sub-contractor changing his design and not originate with the sub-contractor and be the means of achieving the unamended design concept.

Design responsibility

In view of this, one of the purposes of the JCT employer/nominated sub-contractor agreement NSC/W is to provide direct design responsibility from sub-contractor to employer. As well as giving the employer a safeguard, it also protects the architect who would be entirely responsible for the design in the absence of it and the stipulations in his own conditions of appointment (see 'Design warranty', hereafter). The architect should therefore obtain the employer's agreement to this delegation of design, and not dictate to the sub-contractor over design, as distinct from integration.

The agreement is entered into between the employer and the sub-contractor and becomes fully effective when the contractor and sub-contractor enter into their sub-contract, following the architect's nomination instruction. There is no financial consideration offered by either party for entering into this agreement, but each agrees to perform certain of his obligations variously under the main or sub-contract to which the other is not part. (Any doubt over the validity of contractual consideration could be removed by a nominal direct payment from employer to sub-contractor introduced into the agreement.) While they are both obligated to the contractor so to do, they have no other direct obligation to each other and so the agreements give a direct route for redress. These matters relate to elements like the sub-contractor meeting the programme and the employer making direct payments to him when the contractor defaults.

Among them are introduced the provisions about design which, as mentioned, do *not* occur in the main or sub-contract. These regulate very selective aspects and do not delineate the extent to which the sub-contractor has performed or is yet to perform design, or how he communicates his design to the architect for review and integration into the whole scheme for the works. The procedural detail given in the clauses for performance specified work is not paralleled. There is no control of sub-letting of design, presumably because the sub-contractor would seldom contemplate something which might detract severely from his specialism. Nor do they provide for entirely separate reimbursement of the cost of design. It is therefore necessary, firstly, for the architect to set out the requirements over design when the sub-contractor is originally approached about the work. The architect must make clear what is immutable or complete in his own design and what may be amended or developed. He must also set out what information in terms of drawings and specifications is to be provided with or even before the tender and what is to be provided later, remembering that the NSC/T tender says nothing. It is also necessary, secondly, for the sub-contractor to include the cost of design in his tender, even though this is expressed solely as a tender for carrying out the work as part of the main contract which excludes design. He will get his money no other way!

Design warranty

Clause 2.1 of the employer/nominated sub-contractor agreement contains the three elements of design, selection of materials and satisfaction of performance specification already introduced. It also gives the sub-contractor's responsibility over these, as that he 'warrants that he has exercised and will exercise all reasonable skill and care'. The 'reasonable skill and care' level of responsibility has been discussed in Chapter 5 under 'Design liability' as distinctly less than a duty of strict care or one to provide an installation fit for a particular purpose. It is the same level as the architect assumes under his own conditions of appointment, where it is measured not as 'all', but as to 'normal standards of the architect's profession'. These conditions also provide that responsibility for the 'competence, proper execution and performance' of design is lifted from the architect, when the employer has accepted that the sub-contractor will be designing a defined part of the works. They state that the employer 'will hold ... such sub-contractor ... and not the architect, responsible'. This he can do through the present agreement, but not otherwise.

The employer therefore has a clear right to bring an action for negligent breach of care against the sub-contractor. None of these points takes away the responsibility of the architect to 'co-ordinate and integrate' the sub-contractor's design into the wider scheme, as noted above. It is most likely the position that the architect would still be liable for any gross unsuitability of the sub-contractor's design, under this provision of his appointment and also for failure to notice a serious patent error or omission.

Scope of and procedures for design etc

The sub-contractor's work

Clause 2.1 lists three categories of activity for the sub-contractor 'in so far as' they apply. The three categories, which overlap, are given in these terms:

(a) The design of the sub-contract works.
(b) The selection of materials and goods for the sub-contract works.
(c) The satisfaction of any performance specification or requirement included or referred to in the description of the sub-contract works.

The significance of these terms is as discussed in the context of main contracts in Chapters 2, 5 and 17. The necessity to spell out 'how far' the sub-contractor has a function in any category has already been mentioned. Only in the third is it stated that this is to be done within the sub-contract. In that the sub-contractor may write the specification embodied in the sub-contract, he will himself delineate at that stage what is yet to be done by way of design, selection and satisfaction. The clause ends with a disclaimer to the effect that the sub-contractor's obligations under the sub-contract itself are not affected, that is by diminution or addition.

What is *not* done here is to provide anything strictly equivalent to the contractor's proposals under the JCT design and build contract. While the sub-contractor has a responsibility for design etc and while there is no mention of the architect approving anything, or even integrating it into the rest of the project, there is equally no express provision that what the sub-contractor has designed or specified takes any precedence. A skeletal, generalised order of events, when the sub-contractor has major responsibility and the main contract design is firm, would be:

(a) The architect provides tenderers with selected drawings etc for the project and with a brief on what they are to design or specify. This brief may include indicative drawings, elements of materials selection that may or must be followed and any performance specifications.

(b) The tenderers put in their tenders and supporting technical information, showing how they have responded.
(c) The architect considers the tenders and advises the favoured tenderer of his position. He indicates that the design etc are satisfactory as solutions within the total framework, but does not approve calculations or other detail, whatever checking he may have done or had done behind the scenes.
(d) The contractor and the potential sub-contractor settle commercial and attendance details, but do not go into matters of design and specification.
(e) The architect nominates the sub-contractor and the contractor enters into the sub-contract with him.
(f) The sub-contract work proceeds, with the sub-contractor supplying drawings etc for the architect to take into account as necessary in his own design and to pass on to the contractor for his own use and for issue back to the sub-contractor.

This outline assumes that everything proceeds without need for reiteration or revision within a stage, which in practice may be extensive or well-nigh continuous. Items (b), (d) and (e) need no consideration over design. Item (a) may well be expanded by interchanges, so that the more precise the architect can be, the easier life will be all round. Item (c) may also be affected, but with the onus of precision resting on the tenderers and of overall assessment on the architect. Provided that all uncertainties are resolved, all should end well (in relative terms), in that there is also the opportunity to adjust the amounts of tenders before final selection and commitment.

Errors etc

Item (f) in the last list is the difficult one, because there is commitment over design etc and over price, without the same basic philosophy as in a full design and build contract. There the contractor is responsible for his design and price and their relationship, and the ways in which divergences and discrepancies are resolved flow from this and are spelt out accordingly. Here the sub-contractor is responsible over 'reasonable skill and care' in design etc (an expression considered under the next heading), but the rest is left open in the agreement and in the sub-contract. A reasonable set of ground rules for post-contract interpretation would be:

(a) Errors in the architect's requirements, including divergences from other data, not showing on their face are not the sub-contractor's responsibility and the price payable is to be adjusted.
(b) Discrepancies in the architect's requirements should be noticed by the sub-contractor when tendering and drawn to the architect's attention for resolution. If the sub-contractor fails to notice one or makes a reasonable interpretation in his tender he may however still be entitled to an adjustment of payment post-contractually to take account of what is actually required.

This goes back to the *contra proferentem* principle, but may also depend on the strength of evidence as to what the sub-contractor included in his price, and what he should reasonably have noticed.

(c) Discrepancies within the sub-contractor's proposals and divergences between them and the architect's requirements should be noticed by the architect when examining the tender and resolved with the sub-contractor. Again, the architect here may fail or may make a reasonable assumption of intention on the sub-contractor's part. In the absence of misdemeanour by the architect, the sub-contractor may well have to bear the consequences.

(d) Errors in the sub-contractor's proposals not showing on their face are the sub-contractor's responsibility and the price payable is not to be adjusted.

(e) Liability for delay should go with the precedent liability.

In each case, the adjustment or lack of it applies to 'paper corrections' brought in before work is performed and to corrections following fabrication or site work, although the scale of adjustment will vary. There may also be questions of consequential extra work or loss. The boundary in (b) and (c) over what should reasonably be noticed may shift however, when the stage is reached of preparing detailed drawings or comparing all data closely for fabrication etc, rather than in the more global way sufficient for tendering. There is obviously an uncertain, and therefore to the purist unsatisfactory, area in (b) and (c). This may be viewed as necessary in the interests of fairness in very variable situations of shared design and 'quasi-approval' by the architect. If it is considered that contractual certainty must obtain, it becomes necessary to incorporate rules into the documents before tendering: the obvious model for these is in clauses 2.3 and 2.4 of the JCT with contractor's design form, rather than equivalents in the BPF/ACA agreements.

Development and consent

The term 'quasi-approval' needs elaboration. It is coined to emphasise that the architect is not responsible for checking the sub-contractor's design etc in detail, since the sub-contractor gives a warranty to the employer under clause 2.1 of the agreement. As in other situations, the architect may find himself with an unexpected action over liability if the sub-contractor ceases to exist and the employer turns to the architect as still around to be sued. The architect is, however, responsible for seeing that the scheme provides what he requires within the total concept, assuming that the detail is adequate. He should therefore not endorse anything from the sub-contractor as 'approved', but as 'acceptable as fulfilling scheme requirements' or something similar. However, he retains responsibility, for example, for checking such detail as that a pipe and its valves can be adequately housed in a duct which he has designed, unless he has given the schedule of available duct spaces to the sub-contractor. All of this follows through the patterns described in Chapter 5 over the whole works, and particularly as adapted in Chapter 12 over a portion.

Any rules of this nature are far more difficult to formulate, and especially to apply, when the sub-contractor is supplying his scheme within the context of an incompletely designed main scheme. They become partly useless when the sub-contract scheme itself can be only incompletely designed before nomination, because of incompleteness in the main scheme.

In such a situation, the basis for payment is likely to be remeasurement of approximate quantities. Here, the more that the sub-contractor builds or installs, the more he is paid. He thus can afford to provide an additional design margin by using larger structural members, higher capacity services and other elements. The architect should act to curb over-provision, but may then be in danger of 'approving' the design. This is a difficult but sometimes unavoidable situation, quite distinct from when initial approximation is simply due to uncertainty over the areas or lengths to be provided of members of already decided thickness or cross-section. Over-design is not an unknown activity on the part of consultants to the employer: the difference is that *they* are not paid for producing the extra work, only a percentage of it by way of fee!

Subsidiary provisions over design etc

These are in clause 2.2 and 3 and may be epitomised as dealing with early and late design etc which are not used as precise opposites.

Early design

This does not mean pre-tender design here, but design with which the sub-contractor is instructed to proceed between the employer/nominated sub-contractor agreement being entered into and the issue of the nomination instruction. The agreement only becomes fully alive when the nomination instruction is issued, so that this is an interim arrangement. Effectively the period is that following the architect's initial 'approval' of the tender, as an offer to perform work, and includes the time when the contractor and the proposed sub-contractor are tying up details. Indeed, it may consist of little more than that time. It is the period when the sub-contractor is not committed and may withdraw for one of the following reasons, according to the stipulations in the NSC/T tender which may apply:

(a) Learning the identity of the previously unknown contractor, who may be selected after the sub-contractor.
(b) Inability to agree terms with the contractor, after procedures which bring in the architect.
(c) Delay over issuing the nomination instruction.

Each of these includes a time limit, but not a test of reasonableness. If therefore the sub-contractor is instructed and performs design, payment under the present clause cannot then be withheld as a form of penalty for dropping out. The sub-contractor should not proceed with design after dropping out without clearing his position with the architect, even though the agreement does not provide on the one hand that work is to lapse automatically if the sub-contract does not proceed, or on the other run out only after outstanding commitments have been completed.

The positive provision in clause 2.2.1 is simply for the architect to be able to instruct the sub-contractor to 'proceed with ... the designing of ... the Sub-Contract Works'. This allows progress in any waiting period and may be necessary to make design available so that other design or site work is not delayed, or so that materials on long delivery may be ordered or fabricated by the sub-contractor. The clause also allows early ordering or fabrication to be instructed. If an instruction is issued, under clause 2.2.3 the employer is to pay in the case of design 'the amount of any expense reasonably or properly incurred', direct to the sub-contractor. No provisions are laid down about calculation or frequency of payment and retention is not mentioned. In a major case, the architect and sub-contractor should agree relevant terms before the present agreement is finalised.

These payments stop when the nomination instruction is issued, by virtue of clause 2.2.2, so that any balance then unpaid is met in the interim payment made through the main contract. Payments made early are credited in full when settling the sub-contract account under the main contract. These arrangements reflect the fact that the sub-contract sum already includes the cost of all design. The only exception to payment switching over in this way is made in clause 2.2.5. It is in respect of 'any design work properly carried out', but which is not used because of an architect's decision given *before* the nomination instruction is issued. This is still paid for direct, even if performed after nomination. Under the sub-contract itself, because design is not mentioned at all, there is no corresponding separate provision about paying for design work performed, but omitted by an architect's instruction without being used. The sub-contractor should still seek payment in these circumstances, by a reduction in the value of work otherwise omitted.

When the employer pays for design work under clause 2.2.3, it becomes available to him 'for the purposes of the Sub-contract Works but not further or otherwise'. The clause does not say that it becomes his property, as is said over materials in clause 2.2.4. The wording allows the employer, via the architect and contractor, to feed the design back to the sub-contractor for use in the contractually separate sub-contract, or even to a succeeding sub-contractor, if this one drops out before or after nomination. This is to deal with the matter of copyright (see Ch. 5).

Late design

Here this means design performed or supplied later than it should be. Under clause 3.2, the sub-contractor undertakes to supply 'information (including drawings)' early enough for the architect to deal with them and issue 'instructions and drawings' to the contractor on time. If the architect fails to do this, the contractor possibly has an entitlement to an extension of the contract period or to loss and expense payment, or both. The employer has no redress under the main contract against the sub-contractor for failing to perform a design matter, because the contractor (through whom he must act) has no design responsibility. The present clause provides the employer with a cause of action directly against the sub-contractor, to recover lost damages and payments made. The employer cannot deduct them from payments under the agreement, as there are none, and he cannot set them off against main contract payments, and so must act to recover as a debt or by proceedings. Direct payments to the sub-contractor in the contractor's default are not made as payments under the agreement, even though the employer agrees here to 'operate the provisions' of the main contract in question.

Clause 3.1 exonerates the sub-contractor from liability over late design, among other things, until the nomination instruction is issued. This must reasonably mean that the sub-contractor is liable only for delay which he causes after that instruction, and not that retrospective liability may be created.

Part 4

Table of cases and indexes

Table of cases

These cases are referred to or are of interest in relation to discussion in the main text. None relates directly to the contract forms discussed, and few to design and build. Not all were decided on standard forms at all. Different legal decisions based upon the same underlying principles are therefore possible. Specific legal advice is always desirable in any issue of moment.

Where cases are reported in standard law reports, the following abbreviations are used:

AC Appeal Cases
All ER All England Law Reports
BLR Building Law Reports
CA Court of Appeal
CILL Construction Industry Law Letter
ConLR Construction Law Reports
EG Estates Gazette
Ex Exchequer
HL House of Lords
IR Irish Reports
KB King's Bench
Lloyd's Rep Lloyd's Law Reports
QBD Queen's Bench
SJ Solicitor's Journal
WLR Weekly Law Reports

Cases referred to in the text

Further cases for reference

This is an extremely selective set, not referred to in the text, indicating the broad area of subject matter only.

Basis, documentation, programme and quality

Crosby, J. & Sons Ltd v. *Portland Urban District Council* (1967) 5 BLR 121: alternative specifications
East Ham Corporation v. *Bernard Sunley & Sons Ltd* [1966] AC 406, [1965] 3 WLR 1096, [1965] 3 All ER 619: latent defects against 'reasonable inspection'
English Industrial Estates Corporation v. *George Wimpey & Co. Ltd* [1973] 1 Lloyd's Rep 511, CA: special conditions valid for their purpose
Gleeson, M. J. (Contractors) Ltd v. *Hillingdon London Borough Council* [1970] 215 EG 165: standard conditions to prevail

Glenlion Construction Ltd v. *The Guinness Trust* (1987) 39 BLR 89, 11 ConLR 126: contractor's programme with right to finish ahead of contract date, but not necessarily to be supplied with information to permit this

Linden Garden Trust Ltd v. *Lenesta Sludge Disposals Ltd and others; St Martins Corporation Ltd and another* v. *Sir Robert McAlpine & Sons Ltd* (1992) CA, 57 BLR 47, CILL 731: JCT assignment clause permits assignment of benefits, but not obligations

Nevill, H. W. (Sunblest) Ltd v. *William Press & Son Ltd* (1982) 20 BLR 78: defects at completion and consequential loss

Rumbelows v. *AMK and Firesnow Sprinkler Installations Ltd* (1980) 19 BLR 25: exclusion clause by sub-contractor not binding on employer, who was unaware of it

Liquidated damages

Bramhall & Ogden Ltd v. *Sheffield City Council* (1983) 1 ConLR 30: incorrectly entered liquidated damages provisions invalid, although general damages still applicable

Davis Contractors Ltd v. *Fareham Urban District Council* [1956] AC 696, [1956] 2 All ER 145: effects of extreme delay borne between parties

Law v. *Redditch Local Board* (1892) 1 QBD 27, 36 SJ 90: several provisions not in conflict

Disturbance, loss and expense

AMF (International) Ltd v. *Magnet Bowling Ltd and G. Percy Trentham Ltd* [1968] 1 WLR 1028: indemnities voided, but right to damages alive

Hadley v. *Baxendale* (1854) 9 Ex 341: principle of direct loss

James Archdale & Co. Ltd v. *Comservices Ltd* [1954] 1 WLR 459, [1954] 1 All ER 210, 6 BLR 52: indemnity lost by employer's lapse

Minter, F. G. Ltd v. *Welsh Health Technical Services Organisation* (1980) CA, 13 BLR 1: principle of interest or financing charges affirmed, when loss and expense

Saint Line Ltd v. *Richardsons, Westgarth & Co.* [1940] 2 KBD 99: overheads and profit may be allowable

Insolvency and title to goods

Aluminium Industrie Vaassen BV v. *Romalpa Aluminium Ltd* [1976] 2 All ER 552: special clause reversing normal legal position

Humberside County Council v. *Dawber Williamson Roofing Ltd* (1979) 14 BLR 70: standard provisions not securing transfer of title

Payments

Beaufort House Developments Ltd v. *Zimmcor (International) Ltd and Others* (1990) CA, 47 BLR 1: very substantial set-off (enough to cause insolvency) from the employer's side was unjustified when *prima facie* arguable counter claims were outstanding

Croudace Ltd v. *London Borough of Lambeth* (1984) 5 ConLR 70: dilatoriness of employer over payments became breach

Gilbert Ash (Northern) Ltd v. *Modern Engineering (Bristol) Ltd* [1974] AC 689, [1973] 3 WLR 421, [1973] 3 All ER 195, 1 BLR 73: right of common law set-off exists, unless contracted away

Lubenham Fidelities & Investment Co. v. *South Pembrokeshire District Council and Wigley Fox Partnership* (1986) CA, 6 ConLR 85: errors over a certificate, with incorrect deductions for defects and liquidated damages

Rayack Construction Ltd v. *Lampeter Meat Co Ltd* (1979) 12 BLR 30: a trust fund for retention is obligatory upon the employer under JCT forms, whether or not requested.

Taylor, George E. & Co. Ltd v. *G. Percy Trentham Ltd* (1980) QBD, 16 BLR 15: the employer's claim against a nominated sub-contractor for delay was under the collateral agreement and could not be deducted from main contract payments

Tharsis Sulphur & Copper Co. v. *McElroy & Sons* (1978) 3 CA 1040, 5 CR (Ct of Sess Cas) 161: interim payments are provisional and final payment depends on formal instructions

Index of JCT clauses and comparison of section headed format

This index covers all primary divisions of the JCT with contractor's design contract form, showing those places in the main text where substantial discussion of the material is given. Those cases in which no discussion is needed are indicated by an asterisk. Subsidiary references are made in other parts of this book, without entries being made here.

The clause numbers in use at present are given under the ***Current*** column. The anticipated clause numbers of the contract in section headed format are given under the ***Revised*** column to aid tracing clauses in the text in their revised order when the format is published. These numbers correspond at main clause level with the current clauses, but will be subject to sub-division and possibly some rearrangement. They are also given in the headings within the text of related chapters in Part 2 of this book.

8	1.2, 3.11–13	Materials, workmanship	137
9	1.11	Copyright, royalties	93
10	3.4	Person-in-charge	140
11	3.16	Access to works	140
Current	***Revised***		
12	3.6, 3.7, 4.5	Changes and provisional sums	177
13	4.5	Contract sum	127
14	5.3	Value added tax	191
15	1.9	Unfixed materials	139
16	2.6–8	Practical completion	213
17	2.9	Partial possession	216
18	3.1, 3.2	Assignment and sub-contracts	141
20	6.1, 6.2	Injury and indemnity	157
21	6.3	Insurance against injury	157
22	6.6, 6.7	Insurance of works	159
23	2.1, 2.8, 3.14	Possession and completion	145
24	2.5	Damages for non-completion	146
25	2.3, 2.4	Extension of time	147
26	4.7	Loss and expense	153
27	7.1–6	Determination by employer	163
28	7.7–13	Determination by contractor	166
28A	7.14–16	Determination by employer or contractor	168
29	3.10	Work not part of contract	143
30	–	Payments	
	4.2–4	30.1–4	196
	4.1, 4.3, 4.4	30.5–9	219
31	5.4	Statutory tax deduction	191
34	3.18	Antiquities	140
35	4.6	Choice of fluctuations	184
36	4A	Contribution etc fluctuations	189
37	4B	Cost and tax fluctuations	185
38	4C	Price adjustment formulae	189
39	9.1	Arbitration	225
Supplemental	Supplemental	VAT agreement	*
Supplement	Supplement	Sectional completion	216

Subject index

Cross-references and other subsidiary mentions of the following are given here as abbreviations:

BPF	British Property Federation
CDP	Contractor's Designed Portion
CPs	Contractor's proposals
CSA	Contract sum analysis
ERs	Employer's requirements
GC/1	GC/Works/1 contract
ICE	ICE contract
JCT	Joint Contracts Tribunal
NSC	Nominated sub-contractor(s)

Cross-references to multiple first-level entries, such as 'payments', are given without distinction between the several entries. Only in unusual instances are cross-references given to the entries for 'design', which should be consulted regularly for further information.